Human Costs of War

Human Costs of War documents and analyses the direct and indirect toll that war takes on civilians and their livelihoods, taking a human security approach exploring personal, economic, political and community security in Afghanistan, Iraq and Ukraine, in the contexts of the War on Terror and the New Cold War.

The book offers an understanding of war through the recording and comprehension of its civilian casualties and evaluates whether the force used has been proportionate to the threat that prompted it and the concern for human welfare. In the 21st century, the power of the USA has declined, while countries such as China and India become more powerful. The global power balance has been altered in a fundamental way towards a multi-polar world system, with the West no longer able to enforce its policies abroad. Regional and global governance are not assured, and devastating wars have taken a heavy toll in terms of death, poverty and displacement, which feed into the cycle of long-term insecurity. The authors argue that it is important for any conflict to be understood not only in terms of the perpetrators of violence, or of the political and economic reasons behind it, but also in terms of its impact on the civilian population and their security, focusing on conflicts in the Middle East which followed 9/11 and Russia's invasion of Ukraine.

The book will be of interest to academics, the public, the media, security agencies and international organisations. It will be useful for undergraduate and postgraduate students of International Relations, International Law, Security, Politics, Policing, Human Rights, Ethics, Peace Studies, Eastern Europe, American Studies and the Middle East.

Bulent Gokay is Professor of International Relations at Keele University, UK, and a founding member of Iraq Body Count.

Lily Hamourtziadou is Senior Lecturer in Criminology, Security and Policing at Birmingham City University, UK, and a principal researcher of Iraq Body Count.

Innovations in International Affairs
Series Editor: Raffaele Marchetti, *LUISS Guido Carli, Italy*

Innovations in International Affairs aims to provide cutting-edge analyses of controversial trends in international affairs with the intent to innovate our understanding of global politics. Hosting mainstream as well as alternative stances, the series promotes both the re-assessment of traditional topics and the exploration of new aspects.

The series invites both engaged scholars and reflective practitioners, and is committed to bringing non-western voices into current debates.

Innovations in International Affairs is keen to consider new book proposals in the following key areas:

- **Innovative topics**: related to aspects that have remained marginal in scholarly and public debates
- **International crises**: related to the most urgent contemporary phenomena and how to interpret and tackle them
- **World perspectives**: related mostly to non-western points of view

Titles in this series include:

Lessons for Implementing Human Rights from COVID-19
How the Pandemic Has Changed the World
Edited by Oscar Pérez de la Fuente and Jędrzej Skrzypczak

Human Costs of War
21st Century Human (In)Security from 2003 Iraq to 2022 Ukraine
Bulent Gokay and Lily Hamourtziadou

For more information about this series, please visit: www.routledge.com/Innovations-in-International-Affairs/book-series/IIA

Human Costs of War

21st Century Human (In)Security from 2003 Iraq to 2022 Ukraine

Bulent Gokay and Lily Hamourtziadou

Routledge
Taylor & Francis Group

LONDON AND NEW YORK

First published 2025
by Routledge
4 Park Square, Milton Park, Abingdon, Oxon OX14 4RN

and by Routledge
605 Third Avenue, New York, NY 10158

Routledge is an imprint of the Taylor & Francis Group, an informa business

British Library Cataloguing-in-Publication Data
A catalogue record for this book is available from the British Library

ISBN: 978-1-032-54029-0 (hbk)
ISBN: 978-1-032-54036-8 (pbk)
ISBN: 978-1-003-41482-7 (ebk)

DOI: 10.4324/9781003414827

Typeset in Times New Roman
by Newgen Publishing UK

Contents

List of tables   *vii*
Foreword by Paul Rogers   *viii*
Acknowledgements   *xi*

Introduction   1
21st century human (in)security 1
The human security approach 7

1 America's road to the War on Terror   11
September 11: Did all change on that day? 11
Afghanistan 14
Invasion of Iraq, March 2003 19
Project for the New American Century 22
Oil and the War on Terror 25
US world hegemony in decline? 27

2 The human costs of the War on Terror   32
The invasion of Afghanistan 34
The invasion of Iraq 39
Human security 48
Post-traumatic stress disorder 54
International law, casualty recording and human rights 57
Contextualising death and human suffering 61
The mainstream/traditional perspective 66
Towards a human security approach 68

3 The new Cold War: Putin's war in Ukraine, 2014–2022   69
Russia's war in Ukraine 71
The politics of the past 77
What does Putin want? And why did his armies invade Ukraine? 80
*The Izborsky Club (*Izborskii klub*) – 'a call from the past' 83*
War damage and casualties 84

4 War crimes and future wars   88
   War crimes revisited 89
   Recording the casualties 92
   War and ethics: Ukraine's just war 95
   Remote warfare in Ukraine 101
   International justice 107
   Global security dynamics and their impact 114

Conclusion: New Cold War of the 21st century and human security   119

   References   *128*
   Index   *147*

Tables

2.1	Coalition killings of civilians in Afghanistan during 2001–2020	37
2.2	US troops killed at Kabul airport on 26 August 2021	38
2.3	Iraq Body Count incident x025	39
2.4	Iraq Body Count incident j038: nine children playing with Iraqi ordnance, Missan	40
2.5	Iraq Body Count incident d3412	40
2.6	Monthly civilian deaths during 2003–2013	42
2.7	The ISIS years (monthly civilian deaths during 2014–2017)	43
2.8	Iraq Body Count incident a6250	44
2.9	Iraq Body Count incident a5885	44
2.10	Iraq Body Count incident k4435	45
2.11	Bodies found in Baghdad on 1 November 2006	46
2.12	Amina Ayman Ahmed	61
2.13	Iraq Body Count incident a6262	62
2.14	Iraq Body Count incident d12838	62
2.15	Iraq Body Count incident d12801	63
2.16	Iraq Body Count incident j036-i	63
2.17	Iraq Body Count incident a5497	64
2.18	Iraq Body Count incident k001	64
2.19	Iraq Body Count incident k1099	65
2.20	Iraq Body Count incident a6079	65
2.21	Palestinian civilians with the surname Husuna killed in Gaza during 7–26 October	66
3.1	Total estimate of infrastructure damage by industry in monetary terms, as of 1 September 2023	86

Foreword

Paul Rogers

The Human Costs of War adopts a human security approach that places the human impact of war in the context of state actions and also examines state behaviour through the lens of civilian suffering. It concentrates on the period from 2001 with the start of the 'war on terror', especially its multiple conflicts in West and South Asia and much of Africa, through to the war in Ukraine. In doing so, the authors track through the enormous human costs and their changing nature, providing detail and analysis that add substantially to the value of their approach and the need for a much greater public focus on the human costs of war.

The Ukraine war has features common to urban warfare in recent decades, but the war in Gaza combines huge urban destruction and civilian loss of life with elements of counterinsurgency. Both, in different ways, point to patterns of future conflicts that have to be understood if the alternative human security approach is to make progress in diminishing the appalling consequences of war in the coming years.

In an era of uncertainty, there are trends that point to changes in the causes of war in the coming years. One, inevitably, is the onset of climate breakdown. Over the past five years, the pace of climate change has accelerated to the point where the impacts are having increasingly serious consequences across the world from huge wildfires, catastrophic floods, droughts and heat domes to storms and other phenomena. At the same time, the impressive technological developments in decarbonisation mean that the worst excesses of climate breakdown can still be avoided; time is short and there is little sign that sufficient action will be taken in time.

The second trend is the evolving impact of four decades of the neoliberal economic system that has increased the wealth/poverty divide and resulted in billions of people who are marginalised. As climate change transitions to climate breakdown, they will become increasingly desperate for a better life and one of the main consequences of this will be many millions of people needing to escape from parts of the world becoming uninhabitable.

On present trends that will mean that richer and more elite societies will be determined to maintain their own security at the expense of the majority. This will be accomplished, when necessary, by the use of force to protect themselves. It will increasingly be a tendency exacerbated by the world's military-industrial complexes and their emphasis on technologies of control rather than on conflict prevention. That is the third trend, the ready recourse to military responses, and it is firmly embedded in the culture of the world's military-industrial complexes. It will be at the root of responses to a crowded, glowering and divided planet, with elites determined to maintain their wealth and power in an era of numerous revolts from the margins.

Such a dystopic future can be avoided as there are ready responses to all three trends. On the vexed issue of greenhouse gas emissions, the potential for exploiting renewable energy resources has been transformed in the past decade. This is especially true in wind power technology and photovoltaic solar energy systems where costs are typically well below grid parity with fossil carbon energy sources. The potential for more efficient sources of energy storage is also considerable.

On the issue of economic models, there is a growing recognition of the need for a transition from the divisive and unjust neoliberal approach that has been dominant for several decades, as well as a recognition that new models can evolve rapidly without the necessity for sudden revolutionary change. On the third factor, the repeated reliance on military responses to security challenges, there is already innovative thinking in some countries, although there is still much to do.

The world's military-industrial complexes are exceedingly difficult to reform, benefitting from a near-global culture of a ready preparedness for war which itself is all too often a singularly profitable endeavour, yet there are weaknesses in the culture of the complexes that must be exploited. One is the repeated failure of wars to achieve their intended goals, the period covered by this book being a prime example.

Following the 9/11 atrocities, four wars were fought, and all proved disastrous. The war against the Taliban in Afghanistan from 2001 appeared initially to succeed but evolved into a long and bitter insurgency from which the Taliban emerged victorious after two decades. The war in Iraq to terminate the Saddam Hussein regime seemed over within a couple of months by mid-2003 but morphed into a complex pattern of insurgency and confessional violence that lasted close to a decade and still leaves a deeply troubled country.

The 2011 Franco-British campaign to terminate the Gaddafi regime in Libya was over in six months but left a deeply unstable country that also served as a transit route for Islamist paramilitary groups operating across the Sahel and Central Africa. Finally, the intense US-led air war against ISIS in Iraq and Syria from 2014 to 2018 is reported to have killed 60,000 people, including thousands of civilians, but ISIS survives in the Middle East and, along with al-Qaida affiliates, remains active in northern and eastern Africa.

The other weakness, which is the subject of this book, is the terrible human costs of wars. At the time of writing, there are intense wars in Ukraine and Gaza that are covered in the Western media, but other wars elsewhere are largely ignored. Among these are the disastrous civil war in Sudan, the prolonged conflict in the DRC that also involves Rwanda and Uganda, the conflict in Myanmar and the prolonged violence in Afghanistan, repeatedly tipping over into Pakistan.

In all of these, there is the killing, wounding or displacement of ordinary people that may get some attention at times of particular violence, but all too often, it takes second place to the new weapons and tactics that are so beloved by the print and broadcast media. It is in this context that *The Human Costs of War* adds substantially to our knowledge of the often-hidden costs, providing a powerful perspective on the damage done to people and their communities. If we are to avoid an even more fractious and unstable world exacerbated by marginalisation and climate breakdown, such analysis of the full costs of wars is an essential part of moving to a more stable and peaceful world.

Acknowledgements

Our work would not be possible without the men and women who put themselves in harm's way to report and expose a world perpetually in conflict. What they all have in common is a body of work that shows the rest of us who live and work in safety and countries in chaos, engulfed in horrors we cannot imagine or ignore. Their work is inextricably linked to cataclysms, atrocities, war crimes, turmoil and human suffering. War journalists are witnesses to the human cost of war and are themselves frequently part of that human cost. They get caught up in the cross hairs of armies, insurgents, militants and terrorists. During the 2003 invasion of Iraq, 16 Western journalists were dead in the first two weeks; the war would eventually take the lives of nearly 300, most of them Iraqi. War journalists like James Foley have even been executed on camera. While not schooled in the violence of the front lines, they match soldiers in terms of time spent in battlefields, as well as in mental trauma.

Freelance, embedded, reporters, documentarians and cameramen of all nationalities, they all show unparalleled courage and bravery. Too many lose their lives.

Here are a few of them.

Ukraine

French journalist Frédéric Leclerc-Imhoff was on board a humanitarian bus in Ukraine, alongside civilians forced to flee to escape Russian bombs, when he was fatally shot.

Victoria Amelina, Ukrainian writer and journalist, was killed in a Russian missile attack on Kramatorsk.

Rostislav Zhuravlev, a war correspondent working for the Russian state agency RIA Novosti, was hit in a Ukrainian strike in the southern Zaporizhzhia region, dying from his wounds during his evacuation to medical facilities.

Gaza

Hamza al-Dahdouh, Al Jazeera network journalist and cameraman, was driving in a car with other journalists in Gaza, when it was hit by an Israeli drone. Freelance journalist Mustafa Thuraya was also killed.

Iraq

Al Jazeera correspondent Tareq Ayoub was killed when a US warplane bombed Al Jazeera's headquarters in Baghdad during the invasion of Iraq. On the same day, Ayoub was killed, a US tank shelled the Palestine Hotel, home and office to more than 100 unembedded international journalists operating in Baghdad at the time, killing two cameramen, Reuters' Taras Protsyuk and Jose Couso of Spain's Telecinco.

Widad Hussein Ali, a Kurdish journalist reporting for Reuters, was found tortured to death, his body dumped on the side of a road a few hours after his abduction in Dohuk.

Afghanistan

Shah Marai, an Afghan journalist and chief photographer for *Agence France-Presse*'s Kabul Bureau since 1996, died in 2018 from a suicide bomber attack.

Danish Siddiqui, an Indian photojournalist and Pulitzer Prize winner, worked for Reuters. He was killed along with a senior Afghan officer while covering the Afghan Special Forces' attempt to retake an area from the Taliban.

German journalists Karen Fischer and Christian Struwe were killed by the Taliban in Baghlan while conducting research for a documentary.

War scholars owe them a great debt of gratitude.

Introduction

21st century human (in)security

As people worldwide celebrated the new millennium, the world stood on the cusp of what most experts assumed would be a golden age of international peace and prosperity, the inevitable triumph of liberal democracy, guaranteed by an ever-expanding world economy dominated by the West.

> But in an anarchic international system, there cannot be a permanent victory, only a temporary success. Since the end of the Cold War, the international order unravelled quickly and the illusion of the 'end of history' has prevented many from seeing this coming.
>
> (Leoni, 2023: 100)

However, 9/11 and the resulting 'wars on terror', the financial meltdown of 2008 and the Russian invasions of Ukraine followed, leaving most of the world in crisis. Instead of concord, cooperation and equilibrium, this century witnessed renewed great power rivalry, system-wide financial excesses and bursting bubbles centred on power shift from West to East. In addition to this power reshuffle, the 21st century is also witnessing power diffusion to non-state actors, and all these affect the entire world order and lead to instability. The world is currently in a fragile imbalance with a growing number of conflicts and conflict fatalities.

Since 2001, there has been a sharp increase in human insecurity in the world: war and state aggression, terrorism, displacement, poverty and trauma. We continue to see the effects of conflicts in Gaza, Ukraine, Iraq, Afghanistan, Yemen and Syria, to name but a few of the states whose unarmed populations have been impacted. This book makes an important contribution to the monitoring, analysis and documenting of some of the human security crises of the 21st century.

The latest (2024) global explosive violence monitor report from Action on Armed Violence (AOAV), a London-based charity, reveals a disturbing surge in civilian fatalities and incidents of explosive weapon use globally in 2023, with a 122 per cent rise in global civilian fatalities compared to the previous year. The data – collected from English language media sources on explosive violence incidents – highlight an alarming escalation in modern warfare tactics, with a significant impact

DOI: 10.4324/9781003414827-1

on civilian populations, especially in populated areas. The AOAV report reveals a 122 per cent increase in global civilian fatalities from reported explosive violence compared to 2022. It shows:

- A 69 per cent rise in incidents of explosive weapon use.
- Operation Swords of Iron in Gaza contributed substantially to the increase, with 37 per cent of all civilian casualties globally attributed to it.
- Air strikes were responsible for 67 per cent of civilian fatalities.
- 2023 saw the highest number of civilians harmed since AOAV's records started in 2010, with 33,846 civilians killed or injured (AOAV 2024a).

2024 had barely begun, when on 11 January, the Royal Air Force joined coalition forces launching air strikes across Yemen in retaliation against Houthi forces for their attacks targeting ships in the Red Sea. The Houthis responded by pledging to continue their attacks, which they claim are in support of Palestinians as Operation Swords of Iron unfolds. The now ten-year-old conflict in Yemen between the government (supported by a Saudi-led military coalition) and Houthi rebels (supported by Iran) has led to a humanitarian crisis, with widespread hunger, disease and continuous attacks on the unarmed population. Over the past ten years, AOAV has recorded 1,607 incidents of explosive weapons use in Yemen, and 17,197 civilian casualties (7,276 killed).

Air strikes by state actors represent the majority of those incidents, 43 per cent (693), as well as the majority of civilian casualties, 61 per cent (10,450). State-perpetrated air strikes also account for the majority, 70 per cent (5,101) of civilian fatalities. Among the civilians harmed in Yemen over the past decade, 1,422 children were reported as casualties, 59 per cent (845) of whom were killed or injured by state-perpetrated air strikes. Between 2014 and 2023, AOAV has recorded 1,607 incidents of explosive weapons use in Yemen. These incidents resulted in 25,026 total casualties, 69 per cent (17,197) of whom were reported as civilians, including at least 1,422 children. Of these civilian casualties, 42 per cent (7,276) were killed. The most injurious methods used in Yemen over the past decade were air strikes by state actors, with 693 incidents resulting in 10,450 civilian casualties (5,101 killed). State-perpetrated air strikes represent 43 per cent of all incidents recorded in Yemen between 2014 and 2023, as well as 61 per cent of total civilian casualties and 70 per cent of total civilian fatalities. During that time 59 per cent (845) of the children killed or injured in Yemen were harmed by state-perpetrated air strikes. The Saudi-led coalition was reportedly responsible for 38 per cent (614) of total incidents in Yemen in the past ten years, as well as 58 per cent (10,058) of civilian casualties and 68 per cent (4,946) of civilian fatalities. Civilians accounted for 76 per cent of all 13,221 casualties attributed to the Saudi-led coalition at that time. In particular, the 582 air strikes by the Saudi-led coalition recorded by AOAV between 2014 and 2023 caused 9,897 civilian casualties, of whom 4,873 were killed. During that period, 805 children were reportedly killed or injured by Saudi-led coalition air strikes. Over the same time period, ground-launched explosive weapons caused 25 per cent (4,348) of civilian casualties and 18 per cent (1,280) of civilian fatalities, while improvised explosive devices represent 11 per cent (1,919) of total civilian casualties and 9 per cent (665) of total civilian fatalities.

Time after time, we see state aggression impacting the security of civilians even more than acts committed by non-state actors. Since 7 October 2023, Israel has launched 2,000-lb bombs on densely populated neighbourhoods, dropping 6,000 bombs on Gaza in just the first week; in comparison, the USA dropped a little over 7,300 bombs on Afghanistan in all of 2019 (Loveluck, George and Birmbaum, 2023). As of 6 December 2023, according to the Palestinian Ministry of Health (MoH), up to 16,248 Palestinian civilians were killed in Gaza since 7 October, including 7,112 children. This means children account for 44 per cent of fatalities in Gaza so far, as a result of Israel's Operation Swords of Iron. Action on Armed Violence's data shows that 4.4 children are harmed per injurious Israeli air strike in Gaza. So far, Operation Swords of Iron has seen the second-highest rate of children harmed per injurious air strike, and the second-highest percentage of children among the civilian fatalities after Operation Wall Guardian in May 2021 (based on the Office of the High Commissioner for Human Rights [OHCHR] data). Ground-launched attacks have shown an even higher rate of harm to children, with 9.6 children reported harmed per incident (Torelli, 2023).

When explosive weapons are used in populated areas, 90 per cent of casualties are civilians. In total, women and children account for 74 per cent of fatalities in Operation Swords of Iron, at least so far. When compared to data from the OHCHR across three of Israel's most injurious previous operations in Gaza (Operation Pillar of Defense, November 2012; Operation Protective Edge, July–August 2014; and Operation Wall Guardian, May 2021), AOAV concludes that Operation Swords of Iron emerges as the Israeli operation with the second highest percentage of child casualties among civilian fatalities.

Operation Pillar of Defense: 101 civilian fatalities, 33 children (33 per cent) among them.
Operation Protective Edge: 1,462 civilian fatalities, 551 children (38 per cent) among them.
Operation Wall Guardian: 130 civilian fatalities, 67 children (52 per cent) among them.

AOAV has recorded 623 Israeli air strikes in Gaza as part of Operation Swords of Iron (up to 8 December), which have resulted in 9,118 reported civilian casualties (6,400 killed and 2,718 injured). Of the 623 air strikes, 120 have resulted in 533 reported child casualties.

Two such victims were twins Ahmed and Asil Abu Asad, the youngest in their family, killed with their mother on 21 October 2023 in an Israeli air strike. Two days before their deaths, Gaza Martyrs reports, Ahmed went to the barber and styled his hair, saying, 'Even if I die, I want to look neat and good'. Asil was a quiet, kind and gentle girl, who wished to study fashion design. Ahmed loved playing football and dreamed of becoming a football player. Their father encouraged them to study for future success; however, he now regrets this… 'I didn't know that Ahmed wouldn't grow up; I should have let him play as he always wanted' (Gaza Martyrs twitter account, 13 December 2023).

Similar to Gaza, both in Iraq and in Ukraine, there have been staggering numbers of civilian casualties, as the warring parties (a) did not put enough measures in place to provide protection for the civilians, (b) disregarded the safety of vulnerable groups, accepting that there would be 'collateral damage' in the pursuit of their goals, or (c) directly targeted them. The wars' impact has been a lot wider though, affecting all areas of human security. The invasion and occupation of Iraq brought to the country some of the trappings of a liberal democracy – local and national multi-party elections and individual freedoms enshrined in a constitution – but did not ensure the rule of law. Instead, 'a new ethno-sectarian power-sharing system provided opportunities for members of the country's elite to capture the state and plunder its wealth, in many cases with impunity' (Mansour, 2023). In this, they were aided by the legal system, which was under their control, even providing a means for some leaders to repress opponents and increase their own power. The irony, Mansour argues, was that Iraq had a system of governance designed to deliver accountability, but it was not effective. Contemporary accountability mechanisms in the Iraqi bureaucracy do not check members of Iraq's post-2003 elite, many of whom have instead captured these mechanisms, which they then use to dominate the government and resist any genuine reform. 'A similar dynamic hampers public accountability mechanisms (i.e., voting, media, civil society and protests), which instead are all too often taken over by members of the elite to dominate society and set the terms of the public discourse' (Mansour, 2023).

Since the invasion, Iraq has been torn by civil war, terrorism, violations of its borders, attacks from neighbouring states and internal division. Its government is unable or unwilling to combat the misuse of power or to bring perpetrators of severe human rights violations to justice. Human rights defenders, journalists, activists and ordinary citizens continue to suffer from an increased crackdown by the authorities on their right to freedom of opinion and expression. The Iraqi War Crimes Documentation Centre and Alkarama documented the cases of 37 journalists and activists who were murdered from 1 October 2019 to 30 January 2020 (Alkarama, 2022). A report published by the human rights office of the UN Assistance Mission in Iraq (UNAMI) revealed the extent of repression during the 2019–2020 anti-government demonstrations: 'Live ammunition caused most of the 359 deaths of protesters attributed to the security forces at protest sites, with shrapnel and the direct impact of tear gas canisters responsible for 28 other deaths' (UNAMI/OHCHR, 2020: 19).

Executions of men sentenced to death under the anti-terrorism law take place every year in Iraq. Alkarama and the Iraqi War Crimes Documentation Centre highlighted 25,000 men who have been sentenced to death on the basis of the anti-terrorism law. Most of them were arrested and subsequently sentenced to death in the absence of evidence, sometimes solely based on confessions extracted under torture or only on information provided by 'secret informers'. Increasingly, secret prisons are run by militias. Among those secret prisons is a prison in the

Al-Buaitha district of Al-Dora, south of Baghdad, which is under the control of the 'Ashura Brigades'; a prison located in Al-Madaen and placed under the control of Hezbollah; and a prison in Al-Latifah, controlled by the Popular Mobilization. These secret prisons, and many others, are under the control of state-supported militias.

In Ukraine, according to the Small Arms Survey (Hideg, 2023):

- Between 2010 and the end of the 2014–2015 phase of the Russo–Ukrainian war, lethal criminal violence in Ukraine increased by nearly 50 per cent. This raises the possibility of an even steeper future increase in violent crime resulting from the current full-scale war.
- Trust in the police has declined since the winter of 2022–2023, and approximately 14 per cent of Ukrainian civilians now carry some form of weapon for self-defence, which is more than a twofold increase from 2019 and more than a threefold increase from 2011.
- The full-scale Russian invasion of Ukraine initially led to a slight decrease in firearms kept in households as civilians and weapons were mobilised for the war effort. As of the summer of 2023, household ownership levels have gone back up.

Approximately 17.6 million people are in need of humanitarian assistance, according to the UN High Commissioner for Refugees (UNHCR), and over 5 million people are internally displaced. More than 6.2 million refugees from Ukraine have been reported globally as of July 2023 (UNHCR, 2023). Particularly vulnerable groups include older and disabled people who may be unable to flee from high-risk areas. Women and children (approximately 90 per cent of civilians fleeing the war) are at risk of gender-based violence, sexual exploitation and abuse.

Meanwhile, weapons and ammunition are flooding into Ukraine. Ukraine relies almost entirely on an endless stream of equipment and munitions flowing into the country from its partners worth tens of billions of dollars since 2022. On the margins of the NATO summit in Vilnius in July 2023, the leaders of the G7 nations and the EU signed a joint declaration of support for Ukraine based on which 'the signatories must elaborate concrete long-term commitments, including on equipment transfers, military training, intelligence cooperation, and economic support' (Bondar, 2023).

The war in Ukraine has renewed the artificial intelligence (AI) arms race, a race in lethal autonomous weapons, that the War on Terror generated, once again creating moral and legal issues around responsibility, accountability, control, fairness and safety. Autonomous weapons systems (AWS) alter the nature of warfare, putting in question the humanity of our societies. AWS lack compassion and make life-and-death decisions in blatant disrespect of human dignity based on algorithms, daily risking the lives of protected persons. These assassination systems destabilise nations, enable ethnic cleansing by selectively killing particular ethnic groups and do not fit international humanitarian law (IHL) principles. They lead to the exacerbation of regional and international security dilemmas through arms races and the lowering of the threshold for the use of force. Killing remotely has a serious effect on war termination: it makes it much easier to start as well as to continue wars. Ukraine's military has a great demand for drones, especially those that are affordable, swift and lightweight. The Ministry of Digital Transformation launched Army of Drones, a collaborative project involving the General Staff of the Armed Forces, the State Special Communications Service and the fundraising platform UNITED24. The project aims to provide the army with 200,000 drones. 'To achieve this objective, Ukraine is actively promoting domestic production by offering incentives and seeking to nurture the unmanned aviation sector. Army of Drones has entered into approximately 80 contracts with Ukrainian manufacturers' (Bondar, 2023). There has been a significant increase from 30 manufacturers engaged in drone production at the onset of the full-scale invasion to 200 two years on. Ukrainian production is evolving and advancing to include new naval drones, AI-based reconnaissance drones and loitering munitions. The Ukrainian government has committed to the development of an indigenous drone sector by earmarking approximately UAH 43 billion (around $1.2 billion) for the expansion and enhancement of the Army of Drones. The government has also established Brave1, a coordination platform designed to facilitate collaboration among defence industry stakeholders and to provide organisational, informational and financial support for private defence tech projects. 'The Minister of Digital Transformation announced that Ukrainian developers have secured grants exceeding $1,000,000 on the platform. A total of 583 projects were submitted to Brave1, and 305 have already passed the defense examination' (Bondar, 2023). Nineteen ongoing projects worth $315,000 span areas such as drones, robotic systems, electronic warfare, AI tools, cybersecurity, communications and information security management systems.

The war in Ukraine and the War on Terror show that drones play an increasingly prominent role in modern warfare. 'On any single day, up to 20 Russian drones can fly over Yevhen Shovkovplyas's front-line position in Ukraine, and his ageing battery of Soviet-era antiaircraft guns has to stop them', write MacDonald and Sivorka (2023). 'Day and night', write Schecter and Simmons,

> a sound echoes above Khan Younis, the largest city in southern Gaza. The noise isn't the detonation of a bomb dropped by an Israeli jet. It's the low hum of Israeli drones circling overhead. 'They never leave the sky,' Tareq Hajjaj, a freelance Palestinian journalist, said.
>
> (2023)

NBC News visited a Palestinian Authority Fatah party building in the Balata refugee camp, in the West Bank, that had been badly damaged by a drone strike. According to the IDF, the building was a 'hide-out used by terrorists', yet residents said that five people died in the bombing, including a passer-by.

> Another was a 15-year-old boy who died inside it. An interview with his mother that was posted online said he was at his grandfather's house and then went to

the Fatah center before the strike occurred. 'My fate,' his mother said, 'was to become the mother of a martyr'.

(Schecter and Simmons, 2023)

As the bodies of the dead were carried through the streets, drones could be seen circling overhead, monitoring everything, emitting a low hum.

In Afghanistan, 9, 10 and even 11 years after the invasion, UK Special Forces were still killing civilians during night raids. The Independent Inquiry Relating to Afghanistan in 2023 heard, for example, that on 29 November 2010, the SAS had killed Mohammed Ibrahim (55); on 7 February 2011, they killed Sami Ullah (14), Nisah Ahmad (18), Saifullah (22), Izatullah (42), Shamshullah (33) and Haji Wasir (38); on 9 February 2011, they killed Sayeed Mohammed (35), Noor Mohammed (34), Ahmad Shah (32), Abdul Zaheer (23), Mohammed Taher (15) and Ahmad (46); the youngest victims were killed on 6 August 2012. They were Hussain Uzbakzai (24), Ruqqia Mullah Muhammad Haleem (24), Mohammad Wali (26), Mohammed Juma (28), Fazel Mohammed (18), Naik Mohammed (16), Mohammed Tayeb (14) and Ahmed Shah (12).

In Ukraine, victims of Russian aggression include 16-year-old Daria Asiamochkina and her 10-year-old brother Maksym. Both died on 21 July 2023 in Donetsk, when the Russian army shelled their village with artillery (Memorial Platform, 29 July 2023).

Why did these people die? Were their deaths a result of a struggle for power and hegemony? Were they the outcome of clashing civilisations? Or of the ideologies and actions of counter-hegemonic, anti-imperialist resistance movements? Perhaps, in a global context, their deaths were the consequence of a power vacuum and power shift during the crisis of a dying international order. As the New Cold War – in Ukraine and in the Middle East – continues in a changing, multi-polar world, and as its battlefields spread across the globe, a military development and production complex, powerful in determining international security policies, continues to prioritise military responses and state interests to the detriment of human beings. In *Losing Control*, Rogers notes that looking back on the two decades since the 9/11 attacks, those policies and military campaigns have had 'short-term successes followed by long-term failures' (Rogers, 2021: 227). With each 'success' and with each 'failure', the bodies of adults and children pile up.

The human security approach

In the 21st century, the concept of security is no longer interpreted narrowly – it is no longer understood as security of territory from external aggression, protection of national interests or global security from the threat of a nuclear holocaust. The human security approach puts the individual, the citizen and the civilian at the centre of understanding security rather than the state and its borders. The security of people in their daily lives involves a child that did not die of starvation or through lack of medicine. It is a disease that did not spread, a job that was not cut, a dissident who was not silenced and a right that was not violated. Security is indivisible and universal, applying to the wealthy, the poor, Westerners and Easterners, and people of all religions, cultures and races. It is also interdependent, as local insecurities can cross borders and have global implications. Lastly, and most importantly, in the 21st century, security is people-centred: in UN terms, it is 'freedom from fear, freedom from want'. War, poverty, exploitation, imperialism, insurgency and state violence, threaten not only the survival but also the dignity of millions of people. The harm can come as death, illness, starvation, homelessness, bereavement or trauma. It can be harm to personal safety, to basic needs and to freedom of movement, language, culture and self-expression. It can be fear, poverty and inequality and fundamental threats to human security in people's day-to-day existence.

The categories of human security are identified by the UN as economic, food, health, personal, community, political and environmental. In its broader sense, security means freedom from impoverishment, pollution, hunger, homelessness, ill health, lack of healthcare, abuse of power, extremism and oppression (Hamourtziadou and Jackson, 2020). The Commission on Human Security 2001 objectives are:

1   To promote public understanding, engagement and support of human security and its underlying imperatives.
2   To develop the concept of human security as an operational tool for policy formulation and implementation.
3   To propose a concrete programme of action to address critical and pervasive threats to human security.

The Commission on Human Security (CHS) was established in January 2001 in response to the UN Secretary-General's call at the 2000 Millennium Summit for a world 'free of want' and 'free of fear'. On 1 May 2003, Co-Chairs of the CHS, Sadako Ogata and Amartya Sen, presented the Commission's Final Report, Human Security Now, to the UN Secretary-General, Kofi Annan. According to the UN's CHS, human security is:

> to protect the vital core of all human lives in ways that enhance human freedoms and human fulfilment. Human security means protecting fundamental freedoms – freedoms that are the essence of life. It means protecting people from critical (severe) and pervasive (widespread) threats and situations. (...) It means creating political, social, environmental, economic, military and cultural systems that together give people the building blocks of survival, livelihood and dignity.
> (Human Security Now, 2003: 4)

Overall, the definition proposed by the CHS re-conceptualises security by moving away from traditional, state-centric conceptions of security that focused primarily on the safety of states from military aggression to one that concentrates on the security of the individuals, their protection and empowerment; drawing attention to a multitude of threats that cut across different aspects of human life and thus highlighting the interface between security, development and human rights; and promoting a new integrated, coordinated and people-centred approach to advancing peace, security and development within and across nations. As the shift was made from state to person, so the need to account for the human casualties of armed conflict grew. In 2002, the NGO Iraq Body Count (IBC), a human security project, was founded to document civilian deaths in Iraq following the US-led invasion in 2003. Other projects, such as Airwars, a collaborative project aimed at tracking and archiving the international air war against Islamic State and other groups in both Iraq and Syria, and the Syrian Observatory for Human Rights, an information office documenting human rights abuses in Syria since 2011, followed a few years later.

The authors of this book became involved in human security-based International Relations after 2001: Bulent Gokay is a founding member and consultant of IBC and Lily Hamourtziadou its main researcher, casualty recorder and analyst, working with IBC co-founders Hamit Dardagan and John Sloboda. Our involvement with the NGO and research on global security led to publications spanning almost 20 years, most notably *The New American Imperialism: Bush's War on Terror and Blood for Oil* (Fouskas and Gokay, 2005), *The Fall of the US Empire. Global Fault-lines and the Shifting Imperial Order* (Fouskas and Gokay, 2012), *Body Count: The War on Terror and Civilian Deaths in Iraq* (Hamourtziadou, 2021) and *The Ethics of Remote Warfare* (Hamourtziadou, 2024). Through our work, we have inextricably linked human security and war, as human security is threatened the most during acute conflict. Wars such as the one in Iraq, but also Ukraine, Syria, Yemen and Gaza, come with deprivation, disruptions, physical insecurities, use of banned weapons, health crises and inefficiency of healthcare, threats of hunger, crime, ethnic cleansing and even genocide. All civilians face greater insecurity during conflict; it is the experience of living through war. However, different people's experiences differ from conflict to conflict and even within the same conflict. A farmer or a shepherd, for example, will likely face different threats to their security than an urban dweller or a police officer. During conflict, women, men and children experience a variety of threats. Men and boys are more likely to be fighters to kill or to be killed by enemy forces. Women and girls are more likely to be targeted for indentured servitude or sexual violence and exploitation. When ISIS ruled over areas of Iraq, we witnessed both. The war also causes structural changes that threaten people's security. It destroys economies, healthcare systems and agriculture and poisons water supplies. Iraq is the perfect example of how war conditions can impact the lives – and mortality – of people living in conflict-affected areas, as they are not able to access the basic necessities for sustaining life. Loss of life in wartime can be both direct – through violence – and indirect – through the immediate and long-term damage to infrastructure and services. A human security approach to war needs to include the recording, the identification and the contextualisation of all casualties.

The 'New Cold War' of the 21st century as a global security dynamic and strategy, combined with an emphasis on security as a state-centred phenomenon central to the workings of states and military-industrial complexes, rather than one that gives primacy to human lives (human security), has already resulted in several armed conflicts – from full-blown invasions to terrorist attacks. The emphasis on military power and the spending of trillions of dollars, pounds, rubles, hryvnia and euros on preparing for and sustaining wars continue to cause deaths and trauma in civilians and military personnel, widespread displacement and crimes against peace. It is only a human security approach that can address (legally and morally) and bring an end to the human costs of war.

In *Human Costs of War*, we adopt a human security approach, one that places the war's impact on human beings within the context of state actions, as well as examines state behaviour through the lens of civilian suffering, from the War on Terror that was declared in 2001 to the war in Ukraine in 2022. We argue that they are both (like several other conflicts in the Middle East) part of the New Cold War.

The book begins by placing the War on Terror within the context of a global power shift post-Cold War that saw the changing roles of NATO and the EU, as well as the new roles of Russia and America, through political and economic rise/recovery (in Russia's case) and decline (in the case of the USA), impacting international security dynamics. Chapter 2 puts forward the human security approach and examines the ways it shapes our understanding of human suffering. It assesses the impact of war through the lenses of economic, political, military, food, health and personal security. It then looks at the relationship between human security, human rights, international law and the recording of casualties. The psychological impact of war on survivors is also examined by looking at post-traumatic stress disorder for combatants and non-combatants. In Chapter 3, the human security approach is applied to the 21st-century wars between Russia and Ukraine, focusing on the impact on Ukrainian civilians. The chapter assesses civilian harm and argues that it comes as a result of Russian aggression and Western security policies in the New Cold War. The casualties of the latest invasion, we maintain, are to be considered and counted among all those who fell victim to the continuing West–East rivalry and imperial designs from 2001 to the present. The chapter concentrates on Putin's motives and actions, while also looking at the use of politics of the past by the Kremlin, to historically frame its 21st-century conflicts in order to provide justification for policies that have killed tens of thousands. Chapter 4 revisits the concept of war crime in light of the recent trials in Ukraine (and Russia), starting with the conviction of 21-year-old Russian soldier Vadim Shishimarin in May 2022, who was sentenced to life in prison for killing an unarmed Ukrainian civilian by a court in Kyiv. What are war crimes, we ask, and what determines which perpetrators of violence are held accountable and which are not? What does this mean for international law, justice and accountability, as well as for the continuation of acts of aggression against civilians by states and non-state actors? We further ask if there is a type of hard warfare that avoids the commission of war crimes, civilian harm and humanitarian catastrophe. To this end, the chapter evaluates remote warfare and its impact, paying particular attention to war and ethics.

1   America's road to the War on Terror

On 11 September 2001, direct terrorist attacks on the USA unleashed the so-called War on Terror, which began with the US invasion of Afghanistan that same year. It was followed by the invasion of Iraq two years later in what was dubbed 'Operation Iraqi Freedom' – the impact of these invasions is still being felt in these countries more than 20 years later. Enacted on 18 September 2001, *The Authorization for Use of Military Force (AUMF)* allowed the US presidents to 'use all necessary and appropriate force against those nations, organisations, or persons he determines planned, authorised, committed, or aided' the 11 September 2001 attacks, as well as nations that harboured those entities. During this period, US military involvement had spread into an array of other countries across Africa and Asia. The AUMF had been used to justify air and drone strikes and various military operations in Djibouti, Libya, Pakistan, Somalia and Yemen, among others, as well as support for 'partners' in a wide range of countries, including Cameroon, Chad, Eritrea, Georgia, Kosovo, Jordan, Nigeria and the Philippines.

In the intervening years, the narrative of the 'War on Terror' has become one of justifications, explanations and allegations that left devastated and plundered states and peoples in South Asia, the Middle East and Africa that are still struggling for their basic rights. The most obvious legacy of the War on Terror was the lack of democratic accountability and transparency, as well as the millions of refugees, the bodies picked up from the streets of Afghan and Iraqi cities, buried in mass graves, unidentified and unclaimed. The rhetoric and strategy of the War on Terror have shifted across the US presidential administrations, but the lack of democratic accountability persisted for the next 20 years.

September 11: Did all change on that day?

The official Western narrative in September 2001 was that 'everything changed' on the day four airliners were hijacked and nearly 3,000 people murdered. The attack on 11 September 2001 was described as 'a wake-up call against the background of a period of indolence and self-satisfaction' (Kissinger, 2001). It was claimed that 'the new world order is at war and everything is changed utterly – borders, cultures, powers, America, Middle East, Asia, China, Australia' (Rundle, 2001: 2). In the

DOI: 10.4324/9781003414827-2

words of Anatol Lieven, '11 September has ushered in a struggle of civilisation against barbarism' (Lieven, 2001: 15). By many other commentators, we have been told that the day America was attacked 'is a defining moment for humankind' (Held, 2001), and the world will never be the same again (Eichergreen, 2001: 1 and Ash, 2001). According to this interpretation, the US military operations in Afghanistan and the War on Terror, in general, were hastily improvised in less than a month as a direct, and justified, response to the terror attacks on America on September 11. Was the War on Terror, led by the USA, especially the invasions of Afghanistan in 2001 and Iraq two years later, a necessary response to the September 11 terrorist attacks?

When we look at the decisions and debates just before and after September 11, we can see that the decisions that shaped the US-led War on Terror, the military campaigns in Afghanistan and other places, show a remarkable continuity based on an ongoing, pre-September 11 evolution in USA approaches to global security and control. The US administration had been seeking military operations in Afghanistan and other places in the Middle East for some years as a means for achieving global geopolitical goals that had become part of the US global strategy since the end of the Cold War in 1991. Therefore, the causes for the US War on Terror cannot be found by looking only at the events of that day in September 2001 because the roots are much broader and deeper.

To see the whole picture, one has to look at the central shift/change of world history in the late 20th century: the fall of the Soviet Union and its protectorate system in Eastern Europe and the end of the Cold War, 1989–1991. The way the USA initiated its War on Terror was very much a continuation of a policy that started at the end of the Cold War. In other words, there was a significant change in global politics, but this happened not on 11 September 2001, but ten years earlier with the end of the Cold War. In the words of Eric Hobsbawm, the collapse of Soviet power in world politics 'destroyed the …system that had stabilized international relations for some forty years' (Hobsbawm, 1994: 9–11). The dramatic and unprecedented events in Eastern Europe and the Soviet Union during 1989–1991 transformed the geopolitical context from a bipolar to a multipolar world. In particular, it left a vacuum of power and conflicting interests in a zone stretching from Germany in central Europe to China in East Asia – the massive landmass of Eurasia. In the absence of the Soviet Union as an opposing superpower, the USA found itself the master of a new world with unassailable dominance.

The end of the Cold War between the USA and the Soviet-controlled world in the early 1990s produced important shifts in the organisation of the world's economy and politics that seriously challenged the system of global power balance that had existed during the Cold War years. There were major regional powers that were pre-eminent in their own part of Eurasia, but none of them could match the USA in the key dimensions of power – military, economic and technological superiority – that secured global political/military dominance. However, quantitatively superior military power alone could not secure the US political position in this new world, in particular in Eurasia. Rather, it had to require the ability to shape the political and economic context of world politics, which was the goal of the US administration in the post-Cold War era. The way the US administration approached this question created pro-active military interventionism, which was at the core of the War on Terror.

The collapse of the Soviet system in 1991 produced a frenzied race to exploit the rich resources of the region, Central and Eastern Eurasia. The old states of Eastern Europe and the new confederation called the Commonwealth of Independent States (CIS) opened themselves to the big multinational corporations and international banks, triggering a high-stakes game of money and politics, including interventions by the USA, Russian and Chinese governments. At the centre of the post-Cold War world was the rush for control and exploitation of the rich resources of Eurasia – the world's axial super-continent, serving as the decisive geopolitical chessboard for political, military and economic control. Eurasia accounts for 75 per cent of the world's population, 60 per cent of its GNP, and 75 per cent of its energy resources. (Clover, 1999: 9) Collectively, these far overshadow those of any other region or power, including that of the USA, and the lure of the enormous oil and gas reserves of the Caspian Sea basin made the region the focus of fierce competition between multinational corporations and powerful state governments.

In this competition of controlling the region's resources, the USA was the most likely power to succeed. However, it was not as straightforward as it was expected by many. While US military spending exceeded that of the next 13 countries ranked below it, its share of world trade and manufacturing was in decline, substantially, since the 1980s when compared to that of the European Union and the East Asian economic group of Japan, China and the Southeast Asian 'Tigers'. This led to a more activist US foreign policy that since 1989 increasingly relying on its continuing military power. The major US military interventions since 1989 (and they were many) should therefore be viewed not only as justified reactions to 'ethnic cleansing' or 'international terrorism', but opportunistic responses to this post-Cold War geopolitical picture. Andre Gunder Frank wrote about this in June 1999, arguing that this was the post-Cold War strategic-militaristic trend in post-Cold War US foreign policy: 'Washington sees its military might as a trump card that can be employed to prevail over all its rivals in the coming struggle for resources (Frank, 1999: 1). This was a clear strategic use of military power to serve geopolitical and economic ends. 'Washington sees its military might as a trump card that can be employed to prevail over all its rivals in the coming struggle for resources' (A.G. Frank, June 1999 in Gokay and Walker, 2003: 94)

Unimpeded access to affordable energy has always been a paramount strategic interest for the US energy security, both during and in the post-Cold War era. This included controlling firmly the oil and gas resources of Eurasia. Historically, the USA already enjoyed an advantage in the competition for crucial energy and other strategic material supplies in Eurasia, at least since 1945. Its positions in the Middle East with close security relations with the Saudi Arabia, and its sea and air dominance in the Eastern Mediterranean, the Atlantic, the Pacific and the Indian Oceans, provided the USA with a powerful military and political position. In the interest of strengthening its control, the USA was making a determined effort to safeguard this position and cement a permanent role in Eurasia. In the words of former US

Secretary of State Zbigniew Brzezinski in 1997, the immediate task of the USA in 'volatile Eurasia …[is] to ensure that no state or combination of states gains the ability to expel the US or even diminish its decisive role' (Brzezinski, 1997. Securing this dominant role firmly would require, according to the US Energy Secretary, in his interview with Stephen Kinzer, breaking Russian monopoly over oil and gas transportation routes, promoting US energy security through diversified supplies, encouraging the construction of multiple oil and gas pipelines that transit US controlled lands and denying other potential powers' dangerous leverage over the Central Asian oil and natural gas reserves (New York Times, 1998).

The September 11 terrorist attack in New York provided an added incentive to increase this military activist policy to strengthen the US grip over the region as a powerful demonstration of its capacity for political and military control. Indeed, what happened could have come out of what seemed to be the 'wild fantasies' and far-fetched experiments by American strategic analysts as they sought to justify a new, and more active, security role in the post-Cold War world. During the 1990s, considerable effort was devoted to imagining new 'worst case scenarios' that might emerge from a multitude of post-Cold War threats, real or imagined, including possible threats from chemical warfare and biological weapons, to hijacked vehicles and truck bombs, to cyber-terrorism involving jamming 911 services or shutting down electrical, telecommunications and air traffic control systems. The heightened sense of anxiety associated with life in the post 9/11 USA contributed to the politics of fear and pushed the imaginations to an extreme level. President Bush described the War on Terror as a 'clash of civilisations' in which the innocent West is under siege by an inherently violent Islamic evil-doers (Sandole, 2005).

In this imagined 'evil' new world, particular importance was attached to 'rogue states' and 'weapons of mass destruction', and as a consequence, the USA had evolved new hi-tech combat techniques in order to conduct warfare by remote control, where images of destruction could be sanitised and US combat casualties would be radically reduced or eliminated. It was consciously designed to reduce awareness domestically of the actual effects of modern warfare and reduce the costs of war through the use of proxies, such as the Iraqi Kurdish groups in northern Iraq, the KLA in Kosovo and the Northern Alliance in Afghanistan. This strategy had worked to secure quick military victories – in the First Gulf War, in Kosovo, in Afghanistan and Iraq – with minimal US casualties, and helped to get rid of the 'ghost of Vietnam' from American foreign policymaking to some extent. Much of the US War on Terror was designed to be conducted 'remotely', through drones, and through the use of special forces, often through outsourcing as much as they could, but whether it helped to secure long-term and firm US dominance in those regions was another story.

Afghanistan

As mentioned earlier, despite the official claim that the US intervention in Afghanistan was hastily improvised in less than a month, the evidence clearly shows that the decisions shaping the US military campaign in Afghanistan in 2001 reveal a remarkable continuity based on an ongoing pre-September 11 evolution in the US foreign policy. As a matter of fact, the US operations in Afghanistan did not even begin 23 years ago, but in 1979, during the presidency of Jimmy Carter. In 1998, Zbigniew Brzezinski, President Carter's former National Security Advisor, defended US involvement in Afghanistan with its direct support for Islamic extremists fighting against the Soviet army in 1979:

> It was July 3, 1979 that President Carter signed the first directive for secret aid to the opponents of the pro-Soviet regime in Kabul. And that very day, I wrote a note to the president in which I explained to him in my opinion this aid was going to induce a Soviet military intervention. ... That secret operation was an excellent idea. It had the effect of drawing the Russians into the Afghan trap. ... We now have the opportunity of giving to the USSR its Vietnam war. ... What is more important to the history of the world? The Taliban or the collapse of the Soviet empire? Some stirred-up Moslems or the liberation of Central Europe and the end of the Cold War?
>
> (Brzezinski interview, 1998)

The US involvement in Afghanistan began as a CIA-initiated move to arm and unite the Muslims of the country against the occupying Soviet forces (Haslam, 2011: 324). Later under the 'Reagan Doctrine' of 1985, an estimated $3.5 billion was invested in the Afghan war efforts (Misra, 2002: 579). Although Washington stopped its supply of arms to Afghanistan following the Soviet withdrawal in 1989, it did not sever strategic links with the Islamic groups in Afghanistan. By propping up the Taliban, policymakers in Washington thought they could achieve some stable pro-Western regime in Afghanistan with an anti-Shi'a movement in power, which could severely limit Iran's influence in the region.

The details of American involvement in Afghanistan have been well documented in a book written by Ahmed Rashid (2010). The book, *Taliban: The Power of Militant Islam in Afghanistan and Beyond*, provides a perceptive insight into the country and its inhabitants, as well as into the global power contest in the region in a new 'Great Game'. The book narrates the history of the Islamic movement, its origins, its fundamentalism, political and military leaders, internal contradictions, the opium trade and the role of oil resources, as well as foreign interests and involvement in Afghan affairs, in particular from Pakistan, the USA, Russia, Saudi Arabia, Iran and China. Rashid, hardly a wide-eyed radical, has been a Pakistan, Afghanistan and Central Asia correspondent for the *Far Eastern Economic Review* and the *Daily Telegraph* in London. In the book, he argues that the USA was desperately looking for allies and solid bases in energy-rich central Eurasia but was hindered by its own embargo of Iran that began in 1980 and was forced to look for other ways of entering the complex web of Eurasian politics and economics. Sometime in 1994, as Afghanistan tumbled into disarray in the wake of the civil war that followed the Soviet withdrawal in 1989, from the 'jolly bunch' of anti-communist fighters of Afghanistan emerged a highly secretive group of Afghan religious students and scholars, called Taliban. Its declared purpose was to restore peace in the country by confronting crime and corruption, enforcing traditional Islamic law and defending the Islamic character of Afghanistan. The group enjoyed popular support with its promise of security and its religious fervour, and quickly grew into a powerful nationwide movement. By late 1996, the Taliban had seized the capital city, Kabul, and had secured effective control over some two-thirds of the whole of Afghanistan.

Rashid says that in the post-Cold War era it was the key energy interests in using Afghanistan as a major oil transit route that was the main reason why the US administration continued to support this young rebel movement in its search to bring stability to war-torn Afghanistan. Indeed, during the civil war, the Taliban seemed the only force capable of keeping the various ethnic groups in Afghanistan together. Once in power, the Taliban Movement had put a lid on unending banditry, tribal quarrels and sectarian violence, and had disarmed much of the countryside. Some US diplomats who had opened up contact with the Taliban saw them as messianic do-gooders – like born-again Christians from the American Bible Belt (Rashid, 2010: 182). During the Soviet occupation of the country, the US administration actively encouraged, even helped, the recruitment of non-Afghan Islamic mercenaries to fight against the Soviets. For much of the 1990s, the US government supported the Taliban's rise to power, both by encouraging the involvement of US energy companies and tolerating Pakistan and Saudi Arabia, two of its regional allies, in their financial and military backing of the Taliban.

Even though oil and gas were not the only causes of the US involvement in Afghanistan, they occupied a certain place among long-term US policy motivations. Securing control of the vast but land-locked oil and gas reserves of Central Asia was a key US policy, in particular oil and gas reserves of Turkmenistan. The shortest and cheapest export route for Turkmenistan's vast oil and gas reserves was/is through Afghanistan. There were serious negotiations and some advanced planning for both oil and gas pipeline construction by US companies through Afghanistan. The key task of the US administration in 'volatile Eurasia', as described by Zbigniew Brzezinski, was 'to ensure that no state or combination of states gains the ability to expel the US or even diminish its decisive role' (1997: 50–64). Brzezinski also recognised the importance of controlling the flow of energy as the key to power in Eurasia.

> About 75 per cent of the world's people live in Eurasia, and most of the world's physical wealth is there as well, both in its enterprises and underneath its soil. Eurasia accounts for about three-fourths of the world's known energy resources.
> (1997: 31)

This was the case in the 1990s and was still the same in 2001 when the US army invaded the country. Therefore, the causes for the 2001 war in Afghanistan cannot be found by looking only at the September 11 terrorist attacks, without considering this long-term strategic goal. For reasons both of world strategy and control over natural resources, the US administration was determined to safeguard a dominant position in the Eurasian heartland. The 2001 intervention was planned in detail and carefully prepared long before the September 11 terrorist attacks. The attack on September 11 provided an added incentive to the US administration to increase its grip over the region as well as to remind the world of America's capacity for political and military control. A few days before September 11, the US Energy Information Administration reported that

> Afghanistan's significance from an energy standpoint stems from its geographical position as a potential transit route for oil and natural gas exports from central Asia to the Arabian Sea. This potential includes the possible construction of oil and natural gas export pipelines through Afghanistan.
>
> (in Monbiot, 2001)

The hijacked planes that crashed into the World Trade Centre and the Pentagon simply provided an additional rationale for the determination of the US political and military elite to control Afghanistan and the surrounding area. The so-called War on Terror allowed for the US military penetration into areas of the world from which it previously, during the Cold War, had been absent. During the initial stages of the war in Afghanistan, the US military was able to establish 13 new military bases in bordering ex-Soviet states, with Uzbekistan as the first central Asian state to host a permanent military base in early 2002. Shortly thereafter, other bases appeared in Kyrgyzstan and Tajikistan, and the attendant policy and praxis of common military exercises included even distant Kazakhstan. The establishment of this high level of US military presence in the region in early 2000s represented a major advance for the USA. At the same time, these military advances dampened and limited Russian influence in the area. All this also strengthened the position of the USA in relation to China, a power identified since the end of the Cold War as a likely challenger to US hegemony in Eurasia. Within a week of the commencement of the war in Afghanistan, the Bush administration discussed the shape of a post-Taliban Afghan government in reference to developing oil and gas pipelines. On 15 December 2001, the *New York Times* reported that 'the State Department is exploring the potential for post-Taliban energy projects in the region' (Banerjee and Tavernise, 2001).

When the initial fighting concluded, President Bush appointed a former aide to the American oil company UNOCAL, Afghan-born Zalmay Khalilzad, as a special envoy to Afghanistan. This nomination underscored the importance of the economic and financial interests at stake in the US campaign in Afghanistan. Before his ambassadorial appointment, Khalilzad drew up a risk analysis for a proposed gas pipeline from the former Soviet republic of Turkmenistan across Afghanistan and Pakistan to the Indian Ocean. So many business deals, so much oil and natural gas and all those giant multinational corporations with powerful connections to the US state. This is not a paranoid theory, but simply a convergence of political and economic interests travelling under the rubric of 'War on Terror'. It is not conspiracy; it is just business as usual.

US economic interests, driven by oil, had for years taken precedence over any human rights agenda. It was only after 9/11 that the US First Lady Laura Bush emerged overnight as a progressive feminist concerned about the brutal repression of Afghan women under the rule of the Taliban.

> Civilized people throughout the world are speaking out in horror -- not only because our hearts break for the women and children in Afghanistan, but also because in Afghanistan we see the world the terrorists would like to impose on the rest of us. Because of our recent military gains in much of Afghanistan, women are no longer imprisoned in their homes... Yet the terrorists who helped rule that country now plot and plan in many countries. And they must be stopped. The fight against terrorism is also a fight for the rights and dignity of women.
>
> (Bush, L., 2001)

The strict binaries, such as civility vs barbarity and freedom vs oppression of women, moulded American discourses in the War on Terror. As described earlier in this chapter, the USA originally financed the Islamic Mujahedeen upon which the Taliban built its rule as it fought against the pro-Soviet Afghan government of the late 1970s. That war pitted the fundamentalist Mujahedeen against a government that allowed women's access to education and employment. With the fall of this secular government, the Taliban dictatorship was free to enforce strict Islamic rules which included the exclusion of women from all public spaces and education.

From the start, there had been fundamental disagreements on the objectives of the US operation in Afghanistan, both within the US administration and between the USA and its Western allies. Different international and national actors pursued different agendas; they had different priorities and different objectives. At times, some actors pursued conflicting objectives. For some, it was turning Afghanistan into a democracy and bringing cultural change to the country. For others, the main objective was to clear Afghanistan of any terrorist organisations that posed a direct threat to the US interests. In fact, the presence of the USA and other Western forces had remained the basic cause or root of conflict in Afghanistan. Arguably, more lives would have been saved if the USA had left Afghanistan sooner.

The Afghanistan Papers are a set of assessments of the US war in Afghanistan prepared by the Special Inspector General for Afghanistan Reconstruction (SIGAR) and were published by the *Washington Post* on 9 December 2019, following a Freedom of Information Act request. According to these, 2,300 US troops were killed and 20,000 wounded. The Afghans, of course, have suffered far more. Douglas Lute, a retired three-star Army general, advised both the Bush and Obama administrations, stated in his 2015 interview with the SIGAR:

> What are we trying to do here? We didn't have the foggiest notion of what we were undertaking. We never would have tolerated rosy-goal statements if we understood, and this didn't really start happening until Obama. For example, the economy: we stated that our goal is to establish a 'flourishing market economy'. I thought we should have specified a flourishing drug trade – this is the only part of the market that's working. It's really much worse than you think. There

is a fundamental gap of understanding on the front end, overstated objectives, an overreliance on the military, and a lack of understanding of the resources necessary.

(SIGAR, 2021)

Indeed, US-led Western intervention resulted in Afghanistan becoming the world's first true narco-state. Afghanistan's opium production surged from around 180 tonnes in 2001 to more than 3,000 tonnes a year after the invasion, and to more than 8,000 by 2007, and 9,000 by 2018 (93 per cent of the world's illicit heroin supply) (in Gokay, 2022: 8).

When Donald Trump came into office in January 2017, there were 11,000 US troops in Afghanistan, with US force levels having declined from their 2009 to 2011 high point of approximately 100,000 (Lubold and Youssef, 2017). In February 2020, the USA and the Taliban signed a formal agreement in which the USA committed to withdrawing most of its troops, contractors and non-diplomatic US civilians from Afghanistan by mid-July 2020, and a complete withdrawal by the end of April 2021 (White House, 2021a). The final stage of the planned US military withdrawal began on 1 May 2021, and by June, the US Central Command (CENTCOM) reported that almost half of the withdrawal process was complete. Most NATO allies and other US partners withdrew their forces by July. On 8 July, President Biden announced that 'our military mission in Afghanistan will conclude on August 31st' (White House, 2021b).

Invasion of Iraq, March 2003

On 20 March 2003, the US–UK coalition invaded Iraq and toppled Saddam Hussein's regime. According to General Tommy Franks, chief of US Central Command, the objectives of the invasion were:

1 To end the regime of Saddam Hussein.
2 To identify, isolate and eliminate Iraq's weapons of mass destruction.
3 To search for, to capture and to drive out terrorists from that country (Gilmore, 2003).

Of these three objectives, the heaviest emphasis was put on Saddam Hussein's weapons of mass destructions, the so-called WMD, i.e. the claim that the regime of Saddam Hussein had weapons of mass destruction, and that it was developing more of them to the potential advantage of 'terrorist' groups. However, more than 20 years after the launch of Operation Iraqi Freedom, the question of whether the invasion of Iraq was the product of the wilful deception of US and UK governments, or simply wrongful intelligence, is still a matter of debate.

Years later in 2023, Sanam Vakil, deputy director of the Middle East North Africa programme at Chatham House, said that the decision to invade Iraq was a 'huge violation of international law' and that the real objective of the Bush administration was a broader transformational effect in the region. 'We know that the intelligence was manufactured and that [Hussein] didn't have the weapons', she said (Warsi, 2023).

The invasion of Iraq commenced on the night of 19 March 2003, with the 'shock and awe' bombing of Baghdad. Millions around the world sat transfixed in front of their TV screens, watching as CNN displayed bombs and missiles blasting throughout the night. The reports came with the warning that they contained flashing images, and indeed, the sky over Baghdad flashed orange and golden, with horrifying sounds of war and devastation filling our ears. Since the invasion, Iraq Body Count has recorded between 188,000 and 211,000 civilian deaths as a result of violence. The number grows to 300,000 when combatants are included. The US–UK coalition alone has killed over 24,000 Iraqi civilians, and that number grows to 33,000 when we include those killed by Iraqi state forces.

The 2003 invasion of Iraq, led by the George W. Bush administration, destroyed all existing political and economic institutions, aiming to turn Iraq into a 'neo-liberal utopia – a system of American style free market capitalism in the Arab world' (Looney, 2003). When Saddam Hussein's regime was defeated and replaced by the Coalition Provisional Authority (CPA), headed by Paul Bremer, a series of extensive neoliberal measures were quickly introduced. Within a month, 200 Iraqi state-owned companies were privatised, and corporate tax was reduced from 45 per cent to 15 per cent. Foreign firms were also allowed to retain 100 per cent of their Iraqi assets. Furthermore, Iraq's oil revenues, the main source for state expenditure, were put into the UDS-dominated Development Fund for Iraq (DFI). This was held in an account at the Federal Reserve in New York and used for restructuring expenditures. Iraq is an important producer of oil and natural gas. Iraq is estimated to have 112 billion barrels of proven oil reserves, the second largest in the world, but this may be a low estimate due to years of war, intervention and international sanctions, much of Iraq's oil and gas reserves remain unexploited. The economy of Iraq is dominated by the oil sector, which accounted for 90 per cent of government revenues and more than one-third of the country's GDP. Therefore, Iraq's oil income is vital to Iraq's economy and all reconstruction efforts of the country. Keeping the oil income away from the country, in New York at the Federal Reserve, is like sentencing the country to remain poor forever. It is estimated that $150 billion of its oil money was stolen from the country since the US-led invasion of 2003 (Tawfeeq, 2021).

According to US government-sponsored audit reports to the US Congress in January 2005, from the Special Inspector General for Iraq Reconstruction (SIGIR), there was poor government oversight and subcontracting procedures which allowed questionable costs to go undetected. The CPA provided less than adequate controls for approximately $8.8 billion of the DFI and did not establish or implement sufficient managerial, financial and contractual controls to ensure that funds were used in a transparent manner. Consequently, there was no assurance that the funds were used to meet the humanitarian needs of the Iraqi people, the economic reconstruction and repair of Iraq's infrastructure, or any other purposes benefiting the people of Iraq.

In the first year of the occupation, more than 70 American companies and individuals, mainly to well-connected companies, including Halliburton subsidiary Kellogg, Brown & Root (KBR), a company that has become well-known for its ties to Vice President Dick Cheney, won lucrative reconstruction contracts, largely paid for from Iraqi funds. During the same period, Iraqi firms received only 2 per cent of the value of contracts, paid for from Iraqi funds. Among the foreign companies, mainly from the USA and UK, Halliburton subsidiary KBR received 60 per cent of the value of all contracts paid for with Iraqi funds. The audit reports 'paint a picture of disorder and negligence. The audits show that the CPA mismanaged Iraqi funds by keeping poor records and monitoring contracts ineffectively' (Open Society, 2004). Research from 2013 by the *Financial Times* showed that the top ten contractors secured business worth at least $72 billion between them. The CPA deliberately sidelined Iraqi producers in favour of more expensive foreign producers as part of a 'starve then sell policy' (Whyte, 2007). Even after the CPA was dissolved, the US government continued to oversee contracts paid for with Iraqi funds. Professor Dave Whyte of Liverpool University describes the economic performance of the Coalition Provisional Authority as 'one of the most audacious and spectacular crimes of theft in modern history'. 'The suspension of the normal rule of law by the occupying powers', Whyte wrote,

> encouraged Coalition Provisional Authority tolerance of, and participation in, the theft of public funds in Iraq. State-corporate criminality in the case of occupied Iraq must therefore be understood as part of a wider strategy of political and economic domination ... war crimes and offences against Iraqi oil wealth powerfully combined to establish a neo-liberal colonial order.
>
> (Whyte, 2007)

When the war in Iraq officially ended in 2011, with President Barack Obama declaring the withdrawal of the troops, a deeply devastated and traumatised country was left behind, with a bankrupt economy. With hindsight, all that is clear: the invasion of Iraq in 2003 was a colossal and very costly blunder. Nearly 5,000 American soldiers had been killed, many thousands more wounded, hundreds of thousands of Iraqis killed and maimed and more than $2 trillion was spent. Even some leading advocates of the war now call it 'all a big mistake', catastrophe or 'big fat mistake' (Doak, 2023).

In the words of historian Emmanuel Todd, 'The war against Iraq was a military absurdity. The USA won a victory over a country with a barefoot army which had been bled dry. It demonstrated its military omnipotence in Iraq in order to hide its economic weaknesses' (Todd, 2003). The key question, however, is what or who prepared the groundwork for the leaders of the USA and its allies on a profoundly mistaken and criminal course. In seeking to answer such questions about the second Iraq war, in 2003, we need to go back to the early 1990s, when a post-Cold War euphoria was dominant among Western capitals. Francis Fukuyama, a professor of International Political Economy at John Hopkins University and a senior researcher at the Rand Corporation, in his famous 1992 book *The End of*

History and the Last Man, based on an article with the same title written in 1989, argued that mankind had now reached the pinnacle of ideological evolution. In his article and the book, he announced that the great ideological battles between East and West were over, and that Western liberal democracy had triumphed. He claimed that the Western route towards modernism was accepted universally and that non-Western countries would also follow this route. He also believed that the aim of modernisation and advancement oblige all societies to respect Western values (Fukuyama, 1992).

Project for the New American Century

When he wrote 'The End of History?', Francis Fukuyama was a neo-con. He was taught by Allan Bloom,[1] a philosophy professor at the University of Chicago and the author of *The Closing of the American Mind*, and he was a researcher for the Rand Corporation, the notorious think tank for the American military-industrial complex. While Fukuyama's essay and the book were celebrating the victory of the Western liberal democracy, a group of leading neo-cons were taking significant steps to secure this victory of the USA over the rest of the world.

Long before 11 September 2001, even before the first claims about the so-called WMD of Saddam Hussein in the early 1990s, a core group of influential officials and experts in Washington were openly calling for regime change in Iraq. Some never wanted to end the first Gulf War in 1991. In 1992, Paul Wolfowitz, then-undersecretary of defense for policy, supervised the drafting of the *Defense Policy Guidance* document. Wolfowitz had considered the ending of the 1991 First Gulf War against Iraq premature and was in favour of continuing the war for a regime change. In the document, he outlined plans for further military intervention in Iraq as an action necessary to assure 'access to vital raw material, primarily Persian Gulf oil' and to prevent the proliferation of weapons of mass destruction and threats from terrorism even though there was no evidence of Saddam's link to any terrorist groups and weapons of mass destruction (Cirincione, 2003).

The main proponents of this US proactive policy were a relatively small group of well-connected neoconservative ideologues, who later in 1997 launched the Project for the New American Century (PNAC), a neoconservative think tank, with William Kristol and Robert Kagan being the lead members (Borgois, 2020). This group of neoconservatives around the PNAC aimed to set forth a new proactive agenda for post-Cold War foreign and military policy that would ensure that the USA could claim the 21st century as its own, where US military dominance would not only protect US national security and national interests but would also establish a global *Pax Americana*. Frankly and outspoken, they stated that 'the master of Eurasia is the master of the world' and that the USA must secure its mastery in that supercontinent forever, and make sure that no other power challenges its dominant position.

PNAC was an initiative by the New Citizenship Project (NCP), a non-profit organisation which was mostly funded by right-wing organisations and foundations in the USA. Most of these donors were from the corporate world and their main concern was that the USA was losing its economic competitiveness globally. They also had an obsession and fear that 'the liberal intelligentsia' were threatening the capitalistic system. The initial members of PNAC came mainly from the Republicans, but also some other influential people with similar concerns, outside of the Republican Party, joined this new think tank.

The main reason for founding PNAC was that several neoconservatives, including its founder William Kristol, thought that American conservatives did not have an assertive plan for foreign policy to benefit from the opportunities that emerged in the post-Cold War era. So PNAC's agenda was to create this plan, which later has been moulded into documents like the *Rebuilding America's Defenses*. Its aim was explained in PNAC's Statement of Principles as

American foreign and defense policy is adrift. Conservatives have criticized the incoherent policies of the Clinton Administration. They have also resisted isolationist impulses from within their own ranks. But conservatives have not confidently advanced a strategic vision of America's role in the world. They have not set forth guiding principles for American foreign policy. They have allowed differences over tactics to obscure potential agreement on strategic objectives. And they have not fought for a defense budget that would maintain American security and advance American interests in the new century. We aim to change this. We aim to make the case and rally support for American global leadership.
(Micklethwait and Woolridge, 2004: 78)

The main difference between PNAC and the other neoconservative think tanks in the USA is that PNAC mainly focuses on defence and foreign policy, while the others also produce policies on various domestic and economic issues. Francis Fukuyama, too, was a founding member of PNAC. The other signatories of PNAC's founding principles were Donald Rumsfeld, Paul Wolfowitz, Elliott Abrams, Norman Podhoretz, Midge Decter, Zalmay Khalilzad, Gary Bauer, William J. Bennett, Jeb Bush, Dick Cheney, Eliot A. Cohen, Paula Dobriansky, Steve Forbes, Aaron Friedberg, Frank Gaffney, Fred C. Ikle, Donald Kagan, I. Lewis Libby, Dan Quayle, Peter W. Rodman, Stephen P. Rosen, Henry S. Rowen, Vin Weber and George Weigel.[2]

In 1998, the PNAC organised a letter to President Clinton, urging him 'the removal of Saddam Hussein's regime from power' (noi, 1998). In 2000, the group issued an influential report, '*Rebuilding America's Defenses: Strategies, Forces, and Resources For a New Century*'. When George W. Bush won the elections in December 2000, this report became a blueprint for the administration's foreign and defence policies. The report noted,

The U.S. has for decades sought to play a more permanent role in the Gulf regional security. While the unresolved conflict with Iraq provides the immediate justification, the need for a substantial American force presence in the Gulf transcends the issue of the regime of Saddam Hussein.
(ABC News, 2003)

This 90-page report was one of the PNAC's most influential publications, citing the PNAC's 1997 *Statement of Principles, Rebuilding America's Defenses*, it asserted that the USA should 'seek to preserve and extend its position of global leadership' by 'maintaining the pre-eminence of U.S. military forces'. The report's primary authors were Thomas Donnelly and Donald Kagan, and Gary Schmitt was credited as project chairman. It also listed the names of 27 other participants that contributed papers or attended meetings related to the production of the report, six of whom subsequently undertook key defence and foreign policy positions in the Bush administration after 2000 (Buchanan and Gillaume, 2009). It suggested that the preceding decade had been a time of peace and stability, which had provided 'the geopolitical framework for widespread economic growth' and 'the spread of American principles of liberty and democracy'. The report warned that 'no moment in international politics can be frozen in time; even a global Pax Americana' will not preserve itself. Later Robert Kagan said that 'the September 11 attacks shifted and accelerated but did not fundamentally alter a course the USA was already on' (Kagan, 2004: 91).

These views about a more proactive US foreign and security policy were not shared only by the members of the PNAC. In many other places, those who were close to the US leadership expressed similar views since the end of the Cold War in 1991. This belief in Pax Americana was quite popular among a cross-section of the US political spectrum. It is particularly important, in this context, what Zbigniev Brzezinski wrote in his *The Grand Chessboard: American Primacy and Its Geostrategic Imperatives* (1997). He served as US National Security Advisor to President Jimmy Carter from 1977 to 1981. Known for his hawkish foreign policy at a time when the Democratic Party was increasingly dovish, he is a foreign policy realist, hardliner and considered by some to be the Democrats' response to Republican realist Henry Kissinger.

> Although distant, the United States, with its stake in the maintenance of geopolitical pluralism in post-Soviet Eurasia, looms in the background as an increasingly important if indirect player, clearly interested not only in developing the region's resources but also in preventing Russia from exclusively dominating the region's geopolitical space. In so doing, America is not only pursuing its larger Eurasian geostrategic goals but is also representing its own growing economic interest, as well as that of Europe and the Far East, in gaining unlimited access to this hitherto closed area.
>
> (Brzezinski, 1997: 139)

Attained in the course of less than a century, the principal geopolitical manifestation of that hegemony is America's unprecedented role on the Eurasian landmass, hitherto the point of origin of all previous contenders for global power. America is now Eurasia's arbiter, with no major Eurasian issue soluble without America's participation or contrary to America's interests, and fighting for to achieve this goal regime change in a number of Eurasian countries had become the essential method for the US power (Brzezinski, 1997: 193).

Regime change was the dominant thread in US foreign policy from the end of the Cold War, vindicated by this triumph. It was seen as in the national interest that other states should drop authoritarianism for democracy and it went hand in hand with the expansion of NATO to Russian borders. Thus what blinkered American politicians saw as an entirely altruistic strand in their foreign relations was naturally seen by its likely victims and their allies as the vigorous pursuit of US self-interest, backed by the unilateral use of force. And when President Bush attended a summit with Putin in St Petersburg in the summer of 2006, Bush drew attention to the challenges posed by democratic freedoms, especially freedom of the press, in Russia – and then noted that things had gotten much better in Iraq. Putin immediately responded, 'Well, we really would not want the kind of democracy they have in Iraq.' The room filled with applause, and not everyone heard Bush's response: 'Just wait, it's coming.' What Bush had in mind was increased stability in Iraq, but it sounded more ominous: you'll see, democracy will be brought to you as well...

(Haslam, 2022)

Oil and the War on Terror

Well before the USA and UK led the invasion of Iraq in March 2003, the two countries were under suspicion about their intentions for Iraq's oil. When asked on 6 February 2003, one month before the actual invasion, in a BBC Newsnight interview whether the forthcoming war in Iraq was about oil, Tony Blair responded:

Let me just deal with the oil thing ... the oil conspiracy theory is honestly one of the most absurd when you analyse it. The fact is that, if the oil that Iraq were our concern, I mean we could probably cut a deal with Saddam tomorrow in relation to the oil. It's not the oil that is the issue, it is the weapons.

(Gokay, 2016)

But then again, the commander of the USCENTCOM during the war, General John Abizaid, during a round-table discussion on 'the Fight for Oil, Water and a Healthy Planet' at Stanford University in October 2007, stated that: 'Of course [the Iraq war] is about oil, we can't deny that' (Stanford, 2007). Former Federal Reserve Chairman Alan Greenspan agreed, writing in his memoir, 'I am saddened that it is politically inconvenient to acknowledge what everyone knows: the Iraq war is largely about oil' (Juhasz, 2013).

Yet reducing the war in Iraq to this motive alone would be too simplistic, and as the Chilcot Report of the Iraq Enquiry of 2016 makes plain, the explanations for the war are highly complex. Still, some factors are more significant than others (Chilcot Report, 2016). Oil was not the only goal of the Iraq war, but it was certainly a central one, as many leading US military and political figures have confirmed in the years following the 2003 invasion. Today, more than 20 years after the war, it seems the majority view among observers and researchers, including many high-ranking

US officials, is that the oil imperative was a central factor as part of a much more complex system of factors that drove the 2003 invasion of Iraq.

As for Britain, which enthusiastically supported the US-led war in Iraq in sharp contrast to most of the USA's European allies, the partnership with the USA wasn't simply a manoeuvre for influence on the global stage, and nor was it the fruit of imperial delusion. Britain also had a clear material interest in being awarded a share of the spoils in the form of trade, contracts and access to markets and natural resources. This is not a matter of supposition or conjecture. In October 2002, five months before the invasion of Iraq, the then-trade minister, Baroness Symons, told BP executives that the British government was working to secure a good share of Iraq's oil and gas reserves for British energy corporations as a reward for Blair's strong military commitment to US operations for regime change in Iraq (Bignell, McSmith and Brown, 2011).

Confidential minutes released as a result of a Freedom of Information request by Greg Muttitt (co-director of the campaign group *Platform*) reveal that in several meetings with BP and Shell executives, detailed plans were drafted by the British government to exploit oil opportunities in post-Saddam Iraq (Gokay, 2016).

Many consider all this as evidence of a conspiracy, but that is too reductive. These events are part of a larger balance of global economic and political mechanisms – a tangle of political and economic interests converging under the rubric of 'regime change'. Vested interests representing energy, weapons and influential segments of the media and communications industries are entrenched in key sectors of Western governments. These interests are concerned with maintaining their privileged position, and key elements of the USA and British elite respond directly.

> The end of the Cold War … has ushered in a greatly transformed international landscape. Instead of a pacific era of peace and political harmony, the world, and particularly the United States, has been confronted with a menacing challenge of rogue regimes whose propensity for violence is matched by their intentions to disrupt regional stability, contribute to outlaw behavior worldwide, or to possess weapons of mass destruction. Ruthless rogues also endanger American interests and citizens by their active or passive sponsorship of terrorism.
>
> (Hoover Inst, 1999)

If we overlay a map of claimed 'terrorist sanctuaries' and 'rogue states' with one showing the world's principal energy resources, we find considerable congruence between the two. For instance, discussions within the US Energy Information Administration documented Afghanistan's strategic 'geographical position as a potential transit route for oil and natural gas exports from Central Asia to the Arabian Sea' only days long before the 9/11 terrorist attacks and expressed concerns about Taliban regime harbouring terrorist groups in Afghanistan. In the case of war in Afghanistan in 2001 and Iraq in 2003, the tragic events of September 11, 2001 appear as more of a justification for advancing already designed plans to secure US geopolitical interests rather than as a direct response to 'international terrorism'. Subsequent events in Afghanistan and Iraq seem to bear this out, with more US investments in 'security' efforts than in 'nation building'. Because oil was, and still remains, the lifeblood of the modern world economy, superpower status requires control of it at every stage – discovery, pumping, refining, transporting and marketing.

Since the end of the Cold War, the USA has waged several ground and air wars in the region – two in Iraq, one in Afghanistan, one in Libya and one in Syria – and is currently threatening more. Each conflict has, of course, its own specific objectives, but there is a common denominator: the need to keep the oil of the Eurasian regions free-flowing, inexpensive and under the firm control of the USA and its close allies. American strategists don't simply want to obtain the oil. If one has money, that is easy. They also want to eliminate all potential competitors, safeguarding the region politically and militarily so that the flow of oil from the Middle East to world markets remains under their direct control. It is this monopoly over oil production and marketing that provides leverage for the continuation of the US global political and economic supremacy, and control over rich oil reserves of the region would allow the USA to hold an economic sword over the heads of competitor economies where policy disputes arise.

US world hegemony in decline?

There is an important thread that links the geopolitical events of the last four decades, from the collapse of the former Yugoslavia to the post-Yugoslav wars, to the numerous political and economic crises in the post-Soviet space, and to America's 'War on Terrorism' in Afghanistan and its 'Operation Iraqi Freedom'. While there were different local and regional factors and immediate flashpoints in all these different geographies, what links all these events to each other is the American response to the collapse of the Soviet Union and the opportunities and challenges that emerged in these strategic geographies in Eurasia. As Andre Gunder Frank wrote in 1999, 'Washington sees its military might as a trump card that can be employed to prevail over all its rivals in the coming struggle for resources' [in the world's axial supercontinent] (Frank, 1999: 1). The US hegemony globally rests on its ability to control the sources of and transport routes for crucial energy and other strategic materials needed by the USA and other leading industrial powers. This control cannot be reduced solely to matters of economic prosperity, even though it represents a part. Above all, the control of the energy and other strategic materials in the US foreign policy is a strategic one, which mainly concerns exercising global power, a central part of US global hegemony. When a hegemonic power imposes its political and economic authority over a region, it does so in relation to its allies and its local *protégés*. Gramsci used the term 'hegemony' to signify that the dominant power leads the system in a direction that not only serves the dominant group's interests but is also perceived by subordinate groups as serving a more general interest (Gramsci, 1971: 106–120, 161). Harvey's usage of the term is similar: 'the particular mix of coercion and consent embedded in the exercise of political power' (Harvey, 2005: 36). US ally Japan and West European economies are dependent on oil imports from the Middle East, and

US *protégés* in that region, the oil monarchies, require US protection and military and political support. Through its influence over the oil-rich regimes in the region, the USA has consolidated its strategic presence in the Middle East by effectively controlling the 'global oil spigot'. This seems also an effective way to ward off any competition for the top position in the global hierarchy as all its competitors are heavily dependent on this essential source, oil, coming from the Middle East. This process was analysed by Peter Gowan under the apt title *Contemporary Intra-Core Relations*: 'the empire-state offers a mechanism for managing the world economy and world politics which is sufficiently cognisant of trans-core business interests' (Gowan, 2004: 490). This *Pax Americana*, a historically specific inter-state system, was what Peter Gowan fittingly referred to as the 'protectorate system', a US-centred global 'hub-and-spokes' arrangement (Gowan, 2002).

In the post-Cold War era, the US military has been dominant, and its limitations are minimal. Its spending is more than one-third of the global total and as much as that of the next 11 countries combined ranked beneath it (SIPRI, 2022). Yet the economic power of the USA has been in stagnation since the 1970s and has declined since the end of the Cold War. The world economic landscape is rapidly changing, and a very different world is emerging. In particular, the US share of world trade and manufacturing is substantially less than it was just before the end of the Cold War, and its relative economic strength measured against the EU and the East Asian economic group of China, India and the 'South-East Asian tigers' is similarly in retreat. The persistent use of US military power can therefore be viewed as a reaction to its declining economic power and not merely as a response to the post-Cold War threats of terrorism and rogue states. US leaders see their superior military power as the key weapon that can be employed effectively to prevail over all rivals, and thus to stop this decline.

On 22 March 2003, at the beginning of the US-led war against Iraq, General Tommy Franks, chief commander of the US forces in Iraq, was explaining one of the key objectives of Operation Iraqi Freedom as 'to secure Iraq's oil fields and resources' (CNN.com, 2003). Securing US interests regarding the oil resources of the Middle East is not as simple as just going and militarily capturing key positions of a country. Geopolitical events since 2001 have clearly demonstrated that superior military forces of the USA and its Western allies may take but cannot hold Iraq's, Libya's or other Middle Eastern countries' resources and secure firm US control over them. Far from staving off the downfall of the US economic and financial hegemony, the continuing military aggression and arrogance of the US state may instead push the regional powers to distance themselves from its strategic goals. Member countries of OPEC, for instance, have sharply increased deposits in other currencies, including the euro and the Japanese yen, and placed less in dollars starting from 2001 and the Afghan War. OPEC members cut the proportion of deposits held in dollars from 75 per cent in the third quarter of 2001 to 61.5 per cent. US dollar-denominated deposits fell from 75 per cent of total deposits in the third quarter of 2001 to 61.5 per cent in the last quarter of 2004. During the same period, the share of euro-denominated deposits rose from 12 per cent to 20 per cent (European Central Bank, 2005).

The key to understanding post-9/11 events is not in following the political discourse of the day but in understanding the new realities in the global balance of power that emerged following the collapse of the Soviet system in Eurasia. These are not just the geopolitical issues of providing for security in a multi-polar world but in the restructuring that it has thrust upon the world order generally. In this new world order, old economic competitors return, and new ones emerge to challenge the economic and political *status quo* of American dominance that served so well in a bipolar world. While individuals and states make separate decisions, the pattern that emerges is one dictated by a global economy, competitive trade and global financial activities. In this high-stake game, for the USA, the stakes could not be higher: its economic and military strength rests on its ability to command a leading position and prevent other emerging competitors from challenging its hegemony.

Throughout the modern history of international relations, there has been a recognisable pattern of the rise and fall of great powers that is determined by structural material factors. The essential factor here is how economic growth creates uneven development, which is explained by David Landes:

Economic growth was now also economic struggle – struggle that served to separate the strong from the weak, to discourage some and toughen others, to favour the new ... nations at the expense of the old. Optimism about the future of indefinite progress gave way to uncertainty and a sense of agony.

(Landes, 1969: 240)

In this historical process of great power shifts, some states achieve power but others lose it. This 'differential growth in the power of various states in the system causes a fundamental redistribution of power in the system', writes Robert Gilpin (1981: 13). This analysis is directly related to the definition of global hegemony. The concept of hegemony here refers to global leadership by one power, be it a state or a 'historical bloc' of particular powers, whereby the reproduction of dominance involves the enrolment of other, weaker, less powerful entities and is constituted by varying degrees of consensus, persuasion and, consequently, political legitimacy. Force is not completely absent within such hegemonic arrangements, but in 'normal' times it tends to be the exception rather than the rule with regard to the upholding of the hegemonic order.

Global hegemony is a self-limiting, self-defeating and temporary condition in international affairs. That is because hegemonic power has the responsibility of organising the international system, supplying public goods and intervening when necessary, and these all increase the pressures on and cost for the hegemon. However, there always comes a point when the hegemon finds itself over-committed and unable to bear the cost of maintaining the system any further. The hegemon prioritises domestic obligations over its international commitments or finds it more difficult to stick to its global responsibilities. Either way, hegemony declines and collapses on itself and emerging chaos reigns until another hegemonic state, or a group of states, rises and restores order. When the hegemony of a major power or global superpower is in the declining phase, this affects the entire world order and leads to instability. This not only affects the field of economic power; such economic power shifts will 'have a decisive impact on the military/territorial order' (Kennedy 1987: xxii).

Writing in 2007, Peter Dicken argues,

> In these first years of the new millennium, the global economic map is vastly more complicated than that of only a few decades ago. Although there are clear elements of continuity, dramatic changes have occurred. … there has been a substantial reconfiguration of the global economic map. … Without doubt, then, the most important single global shift of recent times has been the emergence of East Asia – including the truly potential giant, China – as a dynamic growth region.
>
> (Dicken, 2007: 68)

Major shifts of hegemony between states occur infrequently and are rarely peaceful. The transfer of power from the West to the East has been gathering pace since the late 1990s, and Washington think tanks have been publishing thick white papers charting Asia's (China's in particular) rapid progress in microelectronics, nanotech and aerospace and printing gloomy scenarios about what this means for America's global leadership. The American administration considers China to be a potential 'strategic competitor' and has exerted enormous pressure on it since the early 1990s (US National Security Strategy, 2015). Just as the 20th century was the American century, the 21st is becoming the Asian century. Everywhere you turn, there is a sense that the USA is in some form of terminal decline; too divided, incoherent, violent and dysfunctional to sustain its *Pax Americana* (McTauge, 2022).

The shift in geo-economic power from developed Western countries to emerging powers in the Global South has been underway for some time. However, the first decade of the 21st century accelerated this process and revealed the relative weakness of the US-centred developed economies. It has been obvious that the future of the world's global politics and economics will not be determined by Euro-American powers, as a new international system is emerging, shaped by the arrival of the new actors from the Global South. This shift is causing the breakdown of the global political and economic order, with ruling elites in many countries turning towards unconstrained economic and political nationalism and aggressive militarism. The Western world is fast losing its superiority, while new emerging powers aspire to a new order of global politics. However, they are not, yet, able to impose their will on various regional and global conflicts in the world. The West lacks the means to back up its policies in the Middle East, Africa, Ukraine or Southeast Asia. The new emerging powers, on the other hand, are aspiring to a new order of global politics, but they are not yet in a position to impose their will upon various regional and global conflicts in the world.

How do declining great powers behave? Evidence suggests they do not behave well, they do not accept their declining status peacefully. As they decline in their relative power, they are likely to become more aggressive than before. Since the beginning of this century, there has been a sharp increase in the determination of

US leaders to use their military power aggressively. The War on Terror, a sharp expansion in military activity and aggression, may appear to be an increase in power, but indeed it was the opposite: it could not mask a decline in actual power. Indeed, what was happening since 2001 has been that 'the US is projecting its own internal disintegration onto the whole world' (Todd, 2003).

Notes

1 Bloom played a central role in facilitating neoconservatism establish alliances with various other conservative movements in the post-Cold War context. His influence in this regard was apparent in the neoconservative polemics against 'political correctness' and 'postmodernism' in the early 1990s.
2 PNAC Statement of Principles, 3 June 1997. 1 February 2005.

2   The human costs of the War on Terror

During the USA's War on Terror, hundreds of thousands of people have lost their lives in unimaginable violence in Afghanistan, Iraq, Syria, Libya, Pakistan, Somalia and Yemen. Millions of civilians were targeted in the course of mass bombing campaigns, and all this directly or indirectly caused disease, suffering, destruction of homes and hundreds of thousands of deaths, which were repeatedly described as 'collateral damage' (Griffiths, 2015). At the same time, Americans convinced themselves that their hegemony was alive and well – and benign (Dobbs and Goshko, 1996: 25).

Based on official US military data, the armed forces of the USA carried out a minimum of 91,340 air strikes during the 20 years of the War on Terror, 2001–2021. Peaks were seen during the invasion of Iraq in 2003 when the USA declared 18,695 strike sorties. The campaign against ISIS also saw a sustained peak, with more than 9,000 strikes a year from 2015 to 2017 (Piper and Dyke, 2021). As many as 48,308 civilians were likely directly killed by US strikes since 2001, with 2017 being the deadliest year, when 19,623 civilians were killed in Iraq and Syria, according to research carried out by Airwars. The second deadliest year was 2003 when Iraq was invaded by the US–UK coalition. The vast majority (97 per cent) of the reported civilian deaths from US wars since 9/11 occurred in the invasions and occupations of Afghanistan and Iraq, as well as in the campaign against the Islamic State in Iraq and Syria.

In 2011, at the peak of its 20-year occupation, the USA had around 100,000 troops in Afghanistan (BBC, 2017), and in Iraq troop numbers peaked at 166,000 in 2007 (CNN, 2011). In 2014, following the rise of ISIS, the USA and its international partners began an aerial bombing campaign against ISIS in support of allies on the ground.

> Campaigns to force ISIS from the Iraqi city of Mosul and the Syrian city of Raqqa in 2016–2017 saw some of the most intense urban fighting since the Second World War. In Raqqa alone, Coalition strikes reportedly killed at least 1,600 civilians. While the Islamists lost their last territorial stronghold in April 2019, the war continues at a low intensity.
>
> (Piper and Dyke, 2021)

DOI: 10.4324/9781003414827-3

In the Iraq and Syria campaign against ISIS, the US-led coalition has accepted killing 1,417 civilians, a figure far lower than Airwars' estimate of at least 8,300 civilian deaths for that war. In 2016, the USA admitted killing between 64 and 116 civilians in Libya, Pakistan, Somalia and Yemen in counter-terrorism operations, between 2009 and 2015 (Woods, 2016). However, as many as 570 civilians have been killed by US actions in those countries, according to the monitoring organisation New America (Bergen, Sterman and Salyk-Virk, 2021).

After 11 September 2001, the USA also launched drone campaigns targeting militant organisations in Pakistan, Somalia and Yemen, which led to significant allegations of civilian harm. In Pakistan and Yemen, the data that have been collected by The Bureau of Investigative Journalism and Airwars show that at least 969 civilians were killed in Pakistan and at least 412 in Yemen. In Somalia, Airwars documented 333 civilian deaths from both suspected and declared US strikes and actions since the conflict began in 2007 (Piper and Dyke, 2021).

According to the Brown University Cost of War programme, nearly 442,000 civilians have been killed by all parties to these conflicts since 2001. The total number of deaths, including combatants, is 940,000 (Watson Institute, Brown University). The War on Terror has become almost a global endeavour. By 2017, the US Department of Defense said it had around 8,000 'special operators' in 80 countries (Garamone, 2017).

The types of conflict fall broadly into three categories:

- Full invasions and occupations of countries – Afghanistan 2001–2021 and Iraq 2003–2009.
- Major bombing campaigns against ISIS – Iraq 2014–2021, Syria 2014–2021 and Libya 2016.
- Targeted US drone and air strike campaigns against militant and terror groups – Somalia 2007–2021, Yemen 2002–2021, Pakistan 2004–2018 and Libya 2014–2019.

This chapter begins by exploring the casualties resulting from the 2001 invasion and occupation of Afghanistan through research carried out by *The Nation*, Airwars and UNAMA. It further queries the morality and legality of this war by asking whether it was a 'just war' and how that affects how we regard its casualties. The chapter also looks at the casualties of the 2003 invasion and occupation of Iraq, using research by Iraq Body Count, Airwars, WikiLeaks, the Correlates of War project, the Uppsala Conflict Data Program and the Global Terrorism Database. It then moves from personal security to three other categories of human security –political, community and economic – assessing different areas of impact, human cost and insecurity resulting from the War on Terror. The mainstream perspective on casualties is then employed to ascertain which human cost is the most important, or which human loss is the greatest. The psychological impact of war on survivors is then examined, by looking at post-traumatic stress disorder and its effects on combatants and non-combatants, after which the relationship between international law, casualty recording and human rights (the legal-moral human cost) is highlighted. The mainstream, state-centric security understanding will be contrasted with that of human security, leading to different conclusions regarding the killing of combatants and non-combatants in war, where non-combatants are (human) collateral damage, while combatants represent a much more valuable state-related cost. The chapter ends with a deliberation on contextualising death and human suffering in war, focusing on the work of the Iraq Body Count.

When it comes to recording deaths, to fully understand the human cost, the appropriate question is *who* died, not merely how many. Like Orio Palmer, each individual killed had a name; each one was a family member; each one had a role in the community. The history of war is often told from the top down, analysing dilemmas facing presidents and prime ministers, charting strategies and tactics, triumphs and failures. Yet the one consequence of all wars is the abrupt ending of life *en masse*. Telling a story of war from the bottom up uncovers patterns of harm, lends weight to advocacy against the use of certain weapons or tactics and illustrates how the acts of the powerful affect the powerless. Meticulous, ongoing casualty recording enables us to see wars through the lens of civilian suffering and loss.

The invasion of Afghanistan

In *Afghanistan. Britain's War in Helmand*, Reynolds writes that 'Britain's decision to deploy military force to Afghanistan followed the attack by al-Qaeda on the USA in 2001 and was part of a multi-national response' (Reynolds, 2016: 11). The aim was to stop the war-ravaged country from remaining a 'safe-haven' for terrorists. The official plan of the invading countries, the US and the UK, was to stop the expansion of al-Qaeda and remove the al-Qaeda-affiliated Taliban from power in Kabul. This was followed by a NATO proposal of a multi-national security oper-ation, under the umbrella of International Stabilisation Assistance Force (ISAF), which would deliver security and stability and allow development to take place. In Helmand, Reynolds claims British forces were welcomed by local people, and it was only criminals who viewed them as a threat. The much-welcomed invading British forces had by the end of the first tour in 2006 'fired 450,000 rifle and machine gun rounds, 7,5000 mortar bombs and 4,000 105-mm light gun shells, illustrating the intensity of the operation' (Reynolds, 2016: 11). There were 3,345 documented civilian deaths directly caused by coalition forces from 2001 to 2006 recorded by *The Nation* (2001–2005) and UNAMA (2006).

As for the USA, Britain's coalition partner, its weapon of choice was the Mark 82 500-lb bomb, which was designed to be guided by the pilot, a nearby obser-vation plane or a spotter on the ground. But there was nothing accurate about the bomb that fell on Bibi Mahru during the invasion.

It killed Gul Ahmad, 40, a Hazara carpet weaver, his second wife Sima, 35, their five daughters and his son by his first wife. Two children living next door were also killed. 'We buried them together in the graveyard. We divided it with sep-arate gravestones but their bodies were all in pieces,' said Arafa, Mr Ahmad's first wife, who was living in another village at the time of the bombing. Taliban officials arrived shortly after the bombing and paid for the funerals.

(McCarthy, 2001)

On 12 November, as the coalition targeted a senior Taliban figure, Jalaluddin Haqqani, the area was bombed three times. At 7:00 pm a 500-lb bomb hit a small group of buildings opposite Haqqani's home, destroying a guard hut where ammunition was stored and a civilian house. Shortly after, a US helicopter fired a rocket blowing the van into the sky. 'Ghulam Ali's sister-in-law Ayesha, 30, was drawing water from a barrel in her front yard nearby when it struck. She was killed instantly by shrapnel' (McCarthy, 2001).

Earlier that day, American planes had targeted a military garrison. Four bombs hit the area.

One landed at the corner of apartment Block 33, where a crowd of children were playing. Nazila, six, was crushed to death by a concrete block. 'She couldn't run away in time,' said her father, Abdul Basir. 'We believed because this was a residential block they wouldn't hit it. We thought they were hitting their targets accurately.' A second landed in the road, a third landed on two houses, killing five people, including a 15-year-old girl.

(McCarthy, 2001)

As we discussed in detail in Chapter 1, long before the terrorist attacks of 2001, US neoconservatives associated with the Project for a New American Century (PNAC) were proclaiming that the USA should dominate as much of the planet as possible in the 21st century and 'prevent the emergence of any significant rival power' (Smith, 2023: 231). PNAC also endorsed a pre-emptive war against nations that challenged the USA (McCartney, 2015). In the summer of 2001, US diplomats had offered Afghanistan's Taliban 'a carpet of gold' if they surrendered al-Qaeda leader Osama bin Laden and allowed an oil and natural gas pipeline to be built in their country. They threatened 'a carpet of bombs' if they refused (Brisard and Dasquie, 2002: 42–43). The invasion of Afghanistan, Operation Enduring Freedom, began on 7 October 2001. It soon toppled the Taliban government but could not defeat the ensuing insurgency and terrorism. More than 220,000 people died in Afghanistan during the War on Terror in 2013 (International Physicians for the Prevention of Nuclear War, Physicians for Social Responsibility and Physicians for Global Survival, 2015: 15–18).

For over two decades, coalition forces have legitimised 'Operation Enduring Freedom' (2001–2014) and 'Operation Freedom's Sentinel' (2015–2021). The concept of 'Just War' has been used as a theoretical framework to scrutinise to what extent military intervention in Afghanistan has adhered to recognised international principles or has served to obscure military misadventure (Connah, 2020), mainly regarding the right to go to war, *jus ad bellum*. Quick resort to military means ensued when the USA declared the War on Terror, immediately after the September 11 terrorist attacks as an act of self-defence, which the United Nations Security

Council quickly authorised in Resolution 1368 (United Nations Security Council, 2001). In addition, the 107th Congress formed a joint resolution, 'Authorisation for Use of Military Force' (AUMF), giving the president the legal right to use 'all necessary and appropriate force against those nations, organizations, or persons he determines planned, authorized, committed, or aided the terrorist attacks that occurred on September 11, 2001' (Congress, 2001). The perceived threat of further acts of terrorism in the days following 11 September 2001 made military intervention inescapable (Misra, 2004), but also as an act of self-defence and protection of the international community from terrorists. The violent methods involved in countering al-Qaeda and the Taliban regime were seen as 'just purposes' by the Bush Administration and the general public (Cortright, 2011). There was also a conviction that President Bush was chosen 'to lead a global war of good against evil' (Hamourtziadou, 2021) and great support for spreading democracy and freedom to oppressed populations in the Middle East. The declaration of the War on Terror was issued by a legitimate authority, the leader of a sovereign state, strengthening the case for *jus ad bellum*, but jihadists and other anti-west insurgents united against Western crusaders (Kaldor, 2013) with terrible consequences for the unarmed population. Tragically, as the moral philosopher Kant wrote in *Perpetual Peace*, war 'produces more evil men than it takes away' (in Williams, 2012: 32). While nations may 'fight a good war' for a noble cause, making great sacrifices and showing bravery and commitment, rather than cowardice and indifference, they still end up destroying their fellow human beings.

Deadly incidents included a suicide attack in Kabul in May 2017, killing 93 people, a police attack in August 2017, killing 72 people, and a suicide attack in October 2017, killing 74 people (Institute for Economics and Peace, 2018). In May 2020, an attack on a maternity ward of the Dasht-e-Barchi hospital in Kabul killed 24 people, including newborns (BBC, 2020).

In 2018, the USA dropped 6,823 bombs in Afghanistan (Al Jazeera, 2019). In April 2018, an air strike targeting Taliban officials in Kunduz killed 50 people (Gossman, 2018). Three months later air strikes on a residential area in Kunduz killed 14 women and children (UNAMA, 2019). Such attacks do not comply with the just war principle *jus in bello* regarding the conduct of parties involved in armed conflict.

> The number of people who have been wounded or have fallen ill as a result of the conflicts is far higher than those killed, as is the number of civilians who have died indirectly as a result of the destruction of hospitals and infrastructure and environmental contamination, among other war-related problems. Hundreds of thousands of soldiers and contractors have been wounded and are living with disabilities and war-related illnesses. Allied security forces have also suffered significant casualties, as have opposition forces.
>
> (Watson Institute, Human Costs, 2023)

Even if the invasion of Afghanistan was considered a just war, the USA had non-military responses available following the September 11 attacks that would have been far less costly in human lives. They would not have turned states like Afghanistan into laboratories 'in which militant groups ..., have been able to hone their techniques of recruitment and violence' (Watson Institute, Human Costs, 2013).

There are many unknowns when it comes to civilian harm in war. While militaries keep good records of their dead and wounded, it is often left to local communities, civil society and international agencies to keep records of civilian casualties. The organisations monitoring civilian harm apply different methodologies. Some, like *The Nation*, are remote monitors; they gather all information publicly available. *The Nation* relies on an extensive survey of reliable media accounts for its raw data, and when no media account is available, it uses casualty reports of NGOs, human rights organisations and ISAF.

UNAMA employs a different methodology for Afghanistan. Based until 2021 in Kabul, it deployed field researchers in each province to physically investigate sites of alleged civilian harm and to interview witnesses. 'While this approach can lead to more certainty about circumstances and casualty numbers in an individual event, it may also mean that some locally reported cases can be missed' (Piper and Dyke, 2021). UNAMA recorded 55,041 civilians as having been killed or injured in the conflict. These are the casualties that UNAMA was able to verify – the actual figure will be even higher. Before UNAMA, figures for Afghanistan were provided by *The Nation*.

Table 2.1 Coalition killings of civilians in Afghanistan during 2001–2020

3,027	*The Nation*
100	*The Nation*
41	*The Nation*
37	*The Nation*
24	*The Nation*
116	UNAMA
321	UNAMA
552	UNAMA
388	UNAMA
171	UNAMA
262	UNAMA
125	UNAMA
122	UNAMA
101	UNAMA
103	UNAMA
127	UNAMA
154	UNAMA
393	UNAMA
546	UNAMA
89	UNAMA
6,799	

Source: Piper and Dyke, 2021.

In 2009, in order to reduce the number of civilian casualties, General McChrystal, US Army General in Afghanistan at the time, introduced a policy of 'Courageous Restraint'. When Afghanistan was invaded, and in the years that followed the invasion, 'towns and villages had been reduced to rubble by air-to-ground munitions fired from aircraft', leading to deaths and injuries, but also extreme displacement: 'in towns like Sangin the entire civilian population was displaced as the town's infrastructure was razed to the ground' (Green, 2017: 106). The policy was not popular among ISAF troops, and when McChrystal was sacked by President Obama the following year, Courageous Restraint was quietly dropped. During those years, according to statistics released by the Ministry of Defence, 'eight British soldiers a month were dying in Afghanistan' (Green, 2017: 109).

Given that 6,799 civilians were killed in Afghanistan by coalition forces from 2001 to 2020, that means 43,963 Afghan civilians were killed by non-state actors. According to the Watson Institute Costs of War project, 52,893 'opposition fighters' were killed in Afghanistan, 2,324 US military personnel and 457 UK military personnel (Watson Institute, Figures, 2023). While 2021 figures, the year Afghanistan again passed to the hands of the Taliban, do not exist, two notable attacks took place in Kabul, both in late August: a deadly terrorist attack and a drone strike.

On the day of the drone strike, on 29 August, Afghan aid worker Zemari Ahmadi ran errands for his employer, which included picking up a laptop and delivering water. Mr Zemari, employed as an engineer at the US-based non-profit NEI foundation, dropped off some of his colleagues and headed home to an evening meal with his family – a drive he routinely did. For the Ahmadis, that Sunday afternoon was like any other. Cousins, friends and families of different age groups were mingling in the family courtyard. Mistaking the compound for an ISIS safe house, six Reaper drones surrounded it and, despite the clear presence of civilians, at 4:53 PM, a single Hellfire missile was launched, killing seven children and three adults (Savage, Schmitt, Khan, Hill and Koettl, 2022). The attack was conducted by the

Table 2.2 US troops killed at Kabul airport on 26 August 2021

Marine Corps Staff Sgt. Darin T. Hoover	31
Marine Corps Sgt. Johanny Rosariopichardo	25
Marine Corps Sgt. Nicole L. Gee	23
Marine Corps Cpl. Hunter Lopez	22
Marine Corps Cpl. Daegan W. Page	23
Marine Corps Cpl. Humberto A. Sanchez	22
Marine Corps Lance Cpl. David L. Espinoza	20
Marine Corps Lance Cpl. Jared M. Schmitz	20
Marine Corps Lance Cpl. Rylee J. McCollum	20
Marine Corps Lance Cpl. Dylan R. Merola	20
Marine Corps Lance Cpl. Kareem M. Nikoui	20
Navy Hospitalman Maxton W. Soviak	22
Army Staff Sgt. Ryan C. Knauss	23

Source: Alfonso, 2021.

Over-the-Horizon (OTH) Strike Cell group of the US Central Command. Those killed were Zemari Ahmadi (40), his son Zamir (20), Mr Ahmadi's cousin Naser (30) and the following children: Faisal (16), Farzad (10), Arwin (7), Benyamin (6), Malika (3), Somaya (3) and Hayat (2).

A few days earlier, on 26 August, a blast at Kabul airport had killed 169 Afghan civilians and 13 US troops who were securing the airport for the withdrawal of US forces. The bomber detonated among packed crowds at the airport's perimeter as they tried to flee the country. The youngest US service members were 20 years old. The oldest was 31.

The invasion of Iraq

The mass killing of Iraqis began on the night of 19 March 2003, with the US–UK coalition's 'shock and awe' bombing of Baghdad: Operation Iraq Freedom.

Table 2.3 Iraq Body Count incident x025

Incident	x025
Type	Air raids
Deaths recorded	15
Targeted or hit	
Place	Zafraniya industrial area, Baghdad
Date	30 March 2003
Sources	REU 30 Mar
	WP 30 Mar
	KR 04 May

Individuals for whom personal or identifying details were reported.

IBC page	Identifying details (number if more than one)	Age	Sex
x025-xu328	Esmaeel Abbas Hamza	49	Male
x025-vm313	Muhammed Taha Abbas	12	Male
x025-xc114	Abeer Taha Abbas	9	Female
x025-ek239	Muna Taha Abbas	23	Female
x025-su332	Abbas Esmaeel Abbas	7	Male
x025-eu317	Azhar Ali Taher	33	Female
x025-sc118	Kameela Abd Kathem	49	Female
x025-nm170	Sabah Gedan Karbeet	42	Male
x025-ks353	Husham Sabah Eadan	10	Male
x025-ns336	Malek Sabah Eadan	7	Male
x025-hv299	Ali Sabah Eadan	4	Male
x025-xd174	Madeeha Abd Kathem	48	Female
x025-kx357	Sabeha Awad Merdas	58	Female
x025-ux340	Fatema Zaboon Maktoof	27	Female
x025-kw295	Nora Sabah Gadan	14	Female

Source: Iraq Body Count database.

The terror that began that night was to last for decades: terror from the sky, terror on the ground, terror from the foreign soldier, terror from armed groups, terror from one's neighbour. By the time the invasion was completed, some 7,500 Iraqi civilians had been killed in the air strikes. Each death was recorded by Iraq Body Count; among them were 15 adults and children who lost their lives in Baghdad's Zafaraniya area on 30 March.

'A Dossier of Civilian Casualties in Iraq 2003–2003', published in July 2005 by Iraq Body Count, revealed the extent of the killings between 2003 and 2005. During the invasion and in the two years that followed, 24,865 civilians were reported killed – almost half in the capital Baghdad (Iraq Body Count, 2005). Nearly a third of these civilian deaths occurred during the invasion phase before 1 May 2003. US-led forces killed 37 per cent of all civilian victims in the first two years; anti-occupation forces and insurgents killed 9 per cent; post-invasion criminal violence accounted for 36 per cent of all deaths. The remainder were killed by 'unknown agents'. At least a further 42,500 civilians were reported wounded.

The war claimed lives in a variety of ways.

An early terrorist attack took place on 7 August, when a car bomb killed 17 people, including 6 police officers.

Thousands of civilians have been killed each year since the night of 'shock and awe'. At its peak, in 2006, the conflict claimed the lives of 29,027 civilians. At its calmest, in 2023, Iraq Body Count recorded 537 civilian deaths.

Table 2.4 Iraq Body Count incident j038: nine children playing with Iraqi ordnance, Missan

Incident	j038
Type	Unexploded Iraqi rocket
Deaths recorded	9
Place	Missan province
Date	12 May 2003

Source: Iraq Body Count database.

Table 2.5 Iraq Body Count incident d3412

Incident	d3412
Type	Car bomb
Deaths recorded	17
Targeted or hit	Outside the Jordanian Embassy
Place	Jordanian Embassy, west Baghdad
Date	7 August 2003

Source: Iraq Body Count database.

The invasion of Iraq was planned in accordance with the concept of rapid dominance – rapidity and timeliness in application, operational brilliance in execution – and coordinated by a high-value target (HVT) cell in the Pentagon. In attempting to hit Saddam Hussain through HVT, due to the kill chain time (the time between getting the intelligence and the missile impacting) being too long (around 45 minutes), many civilians were hit instead. In *Kill Chain*, Cockburn quotes a Pentagon analyst, who recalls 'The shortest kill chain we managed in the 2003 war was forty-five minutes. That was the strike on the al-Saath restaurant in Baghdad. We thought Saddam was there. He wasn't, but we did kill a bunch of civilians' (Cockburn, 2015: 137–138). Drones were going to shrink the kill chain to zero by waiting for the target to appear and then launching a missile, but without having the time to truly assess the collateral damage, even if the right target was hit. During and following the invasion, precision strikes targeted purported lairs of various Iraqi Commanders, as well as Saddam Hussain, without success. All 50 high-value individuals survived. Not so lucky, according to former Defense Intelligence Agency analyst Marc Garlasco, were the 'couple of hundred civilians, at least' who were killed in the strikes (Cockburn, 2015: 138).

Regulations stipulated that civilians could be killed, but not too many (not more than 30), not without clearance from higher authority. Garlasco explains, 'If you're gonna kill up to twenty-nine people in a strike against Saddam Hussein, that's not a problem. But once you hit that number thirty, we actually had to go to either President Bush, or Secretary of Defense Rumsfeld' (Cockburn, 2015: 139). The clearance needed to risk the lives of over 30 civilians was frequently requested and was never refused.

By the end of 2013 Iraq Body Count had documented 134,571 civilian deaths by the US–UK coalition, al-Qaeda in Iraq, Mahdi Army, Iraqi government forces and a variety of militia and insurgent groups, through air strikes, raids, suicide bombings, car bombs, IEDs, shootings, abductions and executions.

As British and American troops were withdrawing between 2009 and 2011, killing levels dropped to 4,000–5,000 civilians a year, the lowest since Iraq had been invaded. Yet as al-Maliki's democratically elected government of Iraq escalated its attacks against its citizens killing over 1,300 of them, the death toll doubled in 2013, as nearly 10,000 people lost their lives. But the worst was about to come in the form of the Islamic State of Iraq and Syria and the battles that followed.

It could be argued that June 2014 was the bloodiest since the invasion: 4,088 civilians were killed in one month, according to Iraq Body Count records (in March 2003, Operation Iraqi Freedom had killed 3,986 in 12 days). ISIS was and remains unparalleled in its brutality. In Iraq alone, it has killed over 50,000 civilians, including entire families. The violence it brought to the country in 2014 was hardly new, but one type of killing became its 'signature': executions. In the database of Iraq Body Count many such incidents have been documented, like the 13 executed in Mosul in February 2017.

On some days, the executions reached triple figures, such as the 20–21 October 2016 killings that included 61 children, all of which were buried in mass graves.

Table 2.6 Monthly civilian deaths during 2003–2013

	Jan	Feb	Mar	Apr	May	Jun	Jul	Aug	Sep	Oct	Nov	Dec	
2003	3	2	3986	3448	545	597	646	833	566	515	487	524	**12,152**
2004	610	663	1004	1303	655	910	834	878	1042	1033	1676	1129	**11,737**
2005	1222	1297	905	1145	1396	1347	1536	2352	1444	1311	1487	1141	**16,583**
2006	1546	1579	1957	1805	2279	2594	3298	2865	2567	3041	3095	2900	**29,526**
2007	3035	2680	2728	2573	2854	2219	2702	2483	1391	1326	1124	997	**26,112**
2008	861	1093	1669	1317	915	755	640	704	612	594	540	586	**10,286**
2009	372	409	438	590	428	564	431	653	352	441	226	478	**5,382**
2010	267	305	336	385	387	385	488	520	254	315	307	218	**4,167**
2011	389	254	311	289	381	386	308	401	397	366	288	392	**4,162**
2012	531	356	377	392	304	529	469	422	400	290	253	392	**4,622**
2013	357	360	403	545	888	659	1145	1013	1306	1180	870	1126	**9,852**

Source: Iraq Body Count database.

Table 2.7 The ISIS years (monthly civilian deaths during 2014–2017)

2014	1097	972	1029	1037	1100	4088	1580	3340	1474	1738	1436	1327	**20,218**
2015	1490	1625	1105	2013	1295	1355	1845	1991	1445	1297	1021	1096	**17,578**
2016	1374	1258	1459	1192	1276	1405	1280	1375	935	1970	1738	1131	**16,393**
2017	1119	982	1918	1816	1871	1858	1498	597	490	397	346	291	**13,183**

Source: Iraq Body Count database.

Table 2.8 Iraq Body Count incident a6250

Incident	a6250
Type	Gunfire, executed
Deaths recorded	13
Targeted or hit	Members of three families executed for trying to flee to Mosul's left coast, the families were captured near Al-Hawi area, casualties include women and children
Place	Ghazi Street, central Mosul
Date	12 February 2017

Source: Iraq Body Count database.

Table 2.9 Iraq Body Count incident a5885

Incident	a5885
Type	Gunfire, executed
Deaths recorded	215–281
Targeted or hit	Men and children from areas and villages of northwest of Mosul, including al-Zawiyah village, bodies of victims were then transferred to mass graves by a bulldozer
Place	Agriculture College, north Mosul
Date	20 October 2016–21 October 2016

Victims' demographic information

Number killed	Age	Sex
61	Child	Male
220	Adult	Male

Source: Iraq Body Count database.

It was easy to become an ISIS victim; the group executed those accused of a number of infractions: trying to flee, being homosexual, not being sufficiently covered in public (women and girls), shaving or cutting one's hair (men), listening to music, using a mobile phone, playing or watching football, committing adultery, collaborating with enemies of ISIS (such as Iraqi security forces), and doctors were executed for refusing to treat injured ISIS fighters. Methods of execution, depending on the infraction, included being shot, electrocuted, thrown off a building, drowned, set on fire and having one's head crushed between two rocks.

The British Parliament began to debate further intervention in Syria in September 2014 in the form of 'precision strikes', but in Iraq they had already commenced. The air strikes meant for ISIS took the lives of thousands of innocents, like Mohannad Rezzo, a university professor, and his 17-year-old son, Najeeb; his sister-in-law Miyada and her 21-year-old daughter, Tuka. The four family members were killed when a drone strike flattened their home as they slept, on 21 September 2015.

Table 2.10 Iraq Body Count incident k4435

Incident	k4435
Type	Gunfire
Deaths recorded	35
Targeted or hit	Bodies found executed
Place	Baghdad
Date	1 November 2006

Source: Iraq Body Count database.

A source of information regarding civilian deaths in Iraq is the Iraq War Logs, which were released by WikiLeaks in October 2010. They contain 54,910 records compiled by the US military whose numerical fields register 109,032 violent deaths between January 2004 and December 2009. In these casualty records, casualties are divided into four categories: Civilian (66,081 deaths), Host Nation (15,196 deaths), Enemy (23,984 deaths) and Friendly (3,771 deaths). 'Civilians' are defined as Iraqi and other civilians, including foreign contractors; 'Host Nation' includes Iraqi security forces, all official Iraqi government forces from local police to National Guard and the Iraqi army; 'Enemy' are anti-occupation forces, insurgents or militia opposing the US and coalition forces; 'Friendly' are defined as US and other allied (non-Iraqi) military forces.

The Iraq War Logs are a version of the SIGACTS (Significant Activities) reports compiled by the US military during 2004–2009. Based on an analysis of a sample of 860 of the Iraq War Logs, IBC researchers have estimated that they would add around 15,000 previously unrecorded Iraqi civilian deaths to the IBC public record, as well as many deaths of Iraqi security forces and other combatants. One case illustrates how the logs will add a dramatic increase to detailed knowledge of the war, even when not necessarily adding new numbers to the deaths total. IBC entry k4435 simply records 35 bodies of persons found, typically killed execution style, across Baghdad over the course of 1 November 2006. These few data points – the number, city, date and general cause of death – were the highest level of detail that existed in the public domain about these deaths until now. The Iraq War Logs also contain 35 bodies found on the same date in Baghdad but spread across 27 logs specifying a wide range of details, including the precise neighbourhood and time of day where particular bodies were found and, in many cases, the demographics and identities of those killed.

IBC entry k4435 in its original form:

Table 2.11 shows 27 incidents in the logs recording bodies found in Baghdad on 1 November 2006 that will likely replace the single IBC entry above. The columns show the wealth of incident and victim details that will be added to the public record when the logs are fully integrated into the IBC database.

In July 2010, three months before the WikiLeaks revelations, the US military had released its own records of Iraqi fatalities.

Table 2.11 Bodies found in Baghdad on 1 November 2006

Time	#	City	Sub-city	Weapons	Victim details
1:22	1	Baghdad	Route Al Amin, Al Sadr district	Gunfire, executed	
8:13	2	Baghdad	Karkh district	Gunfire, executed	Names, occupation, sects
10:35	1	Baghdad	Route Kamaliyah, Al Sadr district	Tortured, strangled	
10:45	1	Baghdad	Washshash village, Karkh district	Gunfire, executed	Age, gender, sect
12:00	1	Baghdad	Haifa Street, Karkh district	Gunfire, executed	
12:30	1	Baghdad	Al Mammon, Karkh district	Gunfire, executed	Gender
13:00	1	Baghdad	Diyala Bridge area, Al Resafa district	Gunfire, executed	Gender
13:00	4	Baghdad	Bab Al Muadam area, Al Resafa district	Gunfire, executed	Gender
13:45	1	Baghdad	Kadhimiya area, Kadhimiya district	Gunfire, executed	Name, age, gender
14:00	2	Baghdad	Al Mammon, Karkh district	Gunfire, executed	Names, ages, genders, sects
14:18	1	Baghdad	Al Mammon, Karkh district	Gunfire, executed	Age, gender
14:30	1	Baghdad	Route Al Rasheed, Hay Al Amil, Karkh district	Gunfire, executed	Gender
14:40	1	Baghdad	Karkh district	Gunfire, executed	
14:50	1	Baghdad	Al Mammon, Karkh district	Gunfire, executed	Name, age, gender, sect
15:00	1	Baghdad	Route Al Ghazaliya, Karkh district	Gunfire, executed	Age, gender, occupation
15:15	1	Baghdad	Route Al Shames Market, Hay Al Adel, Karkh district	Gunfire	Name, gender
15:30	1	Baghdad	Route Al Mansour, Al Ghazaliya, Karkh district	Gunfire, executed	Gender
16:00	3	Baghdad	Route Al Ghazaliya, Karkh district	Gunfire, executed	
16:20	1	Baghdad	Route New Baghdad, Al Amin, Al Sadr district	Gunfire, executed, tortured	Gender
16:25	1	Baghdad	Karada, Al Resafa district	Gunfire, executed	Gender
16:30	1	Baghdad	Route Al Thawra, Sadr City, Al Sadr district	Gunfire, executed	Gender
17:00	1	Baghdad	Mada'in district	Gunfire, executed	Gender

Table 2.11 (Continued)

Time	#	City	Sub-city	Weapons	Victim details
17:00	1	Baghdad	Route Kamaliyah, Al Sadr district	Gunfire, executed	Gender
17:10	1	Baghdad	Al Sadr district	Gunfire, executed	Gender
17:30	1	Baghdad	Route Al Rashid, Al Saydiya, Karkh district	Gunfire, executed	Gender
18:35	2	Baghdad	Route Al Mansour, Karkh district	Assassinated	Names, genders
18:40	1	Baghdad	Al Mansour area, Karkh district	Assassinated	Name, age, gender, occupation

Source: Iraq Body Count, 2010.

The U.S. military released its most detailed compilation of data on Iraqi casualties during more than four years of the Iraq war, reporting that 63,185 civilians and 13,754 members of the country's security forces were killed from the beginning of 2004 through August 2008 (…) The casualty figures released by the United States are lower than Iraqi government accounts. Iraq's Human Rights Ministry reported last year that 85,694 Iraqis, including military and police personnel, were killed from the beginning of 2004 through October 2008.

(Fadel, 2010)

During the period covered by the US report, 3,592 coalition troops were killed and 30,068 were wounded.

The report, prompted by a Freedom of Information Act request from George Washington University's National Security Archive, was posted on the US Central Command website. The figures do not include the 2003 invasion casualties or its immediate aftermath, while 'a U.S. military spokesman said it was unclear whether insurgent killings were included in the data' (Fadel, 2010).

A controversial 2006 survey in *The Lancet*, a British medical journal, estimated that more than 600,000 Iraqis had died as a result of the war, a figure more than ten times as high as other estimates at the time. Iraq Body Count put the number during that period at 107,235 documented civilian deaths.

The Correlates of War offers casualty figures only up to 2007, and it is not clear how the data have been collected. In the 2003 invasion of Iraq, described and coded as 'Inter-State War #227', 7,000 combatant fatalities are recorded. Regarding the 'Iraqi Resistance', 'Extra-State War #482', 10,800 combatant fatalities are recorded.

The Uppsala Conflict Data Program (UCDP), like IBC, collects data from 'all news reports which contain information about individuals killed or injured' (Uppsala Conflict Data Program, 2023). Its data reveal 7,929 civilian deaths during the 2003 invasion, but there were no recorded civilian deaths by coalition air strikes in its war against the Islamic State during 2014–2018 (a period during which Airwars recorded 13,250 civilian deaths from air/drone strikes). UCDP has documented 39,349 civilian deaths by Islamic State and 2,629 by the Mahdi Army. It gives a total of 126,708 deaths since 1989.

The Global Terrorism Database (GTD), the website states, 'was developed to be a comprehensive, methodologically robust set of longitudinal data on incidents of domestic and international terrorism. Its primary purpose is to enable researchers and analysts to increase understanding of the phenomenon of terrorism'. Information is drawn from publicly available materials: electronic news archives, existing data sets, books, journals and legal documents.

As the name suggests, all data are limited to casualties resulting from terrorist activity. Here are two examples of GTD incidents.

Incident 201508050066

08/05/2015: A suicide bomber detonated an explosives-laden vehicle targeting police and volunteer fighters in Diyala governorate, Iraq. In addition to the assailant, at least 13 security force members were killed and nine members were injured in the blast. No group claimed responsibility; however, sources attributed the incident to Islamic State of Iraq and the Levant (ISIL).

Incident 201508020073

08/02/2015: Assailants executed 15 police officers in front of city hall in Mosul city, Nineveh governorate, Iraq. No group claimed responsibility for the incident; however, sources attributed the attack to Islamic State of Iraq and the Levant (ISIL).

The number of combatants (all parties to the conflict) recorded by IBC since the invasion is in the region of 90,000, a figure that includes the 4,492 US servicemembers killed in Iraq. According to the Watson Institute, the number of US troops who died fighting the wars in Iraq and Afghanistan had passed 7,000 at the end of 2019 (Watson Institute, 2021), while 405 British Armed Forces personnel were killed in Afghanistan and 136 British Armed Forces personnel were killed in Iraq, as a result of hostile action (Ministry of Defence). Among those killed were Paul Farrelly and Phillip Hewett.

Lance Corporal Paul Farrelly was killed by a roadside bomb while on patrol in the Al Jezaizah district of northwest Basra. He died on 28 May 2006 at the age of 27. He served with the 1st Queen's Dragoon Guards (The Welsh Cavalry).

Private Phillip Hewett died in a roadside bomb attack on a patrol of three armoured Snatch Land Rovers. He was driving at the time of the attack and died later of his injuries on 16 July 2005. He served with the 1st Battalion, C Company Staffordshire Regiment. He was 21 years old (BBC, 2016).

Human security

'Security', wrote Mary Kaldor, 'is about confronting extreme vulnerabilities' (Kaldor, 2007: 183). Vulnerabilities and insecurities exist in all areas of human life, and what distinguishes the human security approach from traditional, mainstream, state-based approaches is the primacy of human rights, including economic, social, political and civil rights.

Economic security means having an assured income, but also public safety net measures ensuring income to those unable to obtain one. *Food security* requires adequate access to food, physically and economically. *Health security* means having access to health care and protection against diseases. *Personal security* addresses threats from physical violence. *Community security* is the security individuals get within a group, establishing a sense of belonging and identity rooted in shared practices and values. *Political security* requires the freedom to be governed in a way that respects basic human rights, protected by democratic institutions in which individuals are given a voice.

The US military fired thousands of rounds of depleted uranium (DU) during two high-profile raids on oil trucks in Islamic State-controlled Syria in late 2015, the first confirmed use of this armament since the 2003 Iraq invasion, when it was used hundreds of thousands of times, setting off outrage among local communities, which alleged that its toxic material caused cancer and birth defects.

According to the International Committee of the Red Cross, chemical, biological, radiological and nuclear agents (CBRN) have four key properties in common.

- Toxicity: a measure of the ability of a toxic substance to cause harmful effects or death.
- Latency: the interval between exposure to a CBRN agent and the first signs and symptoms of illness or disease.
- Persistency: the capacity of a CBRN agent to remain capable of causing significant harm for a prolonged period of time.
- Transmissibility: whether an agent can be transmitted from one person to another. The main means of transmission of CBRN agents are cross-contamination and direct physical contact.

In 2014, in a UN report on DU, the Iraqi government expressed 'its deep concern over the harmful effects' of the material. DU weapons, it said, 'constitute a danger to human beings and the environment' (Oakford, 2017). DU is radioactive. It is both a toxic chemical and radiation health hazard when inside the body. If ingested or inhaled, it targets organs such as the kidneys and lungs. DU – a waste product of nuclear power generation – is effective in anti-tank projectiles. The radioactive metal reaches high temperatures on impact with tank armour, melting it into minute particles that are carried on the wind as dust. Scientists argue that this radioactive dust contaminates air, water and soil and has harmful consequences for human health: high incidences of cancer, leukaemia and severe birth defects.

Iraqi scientists with the Ministry of Environment and Ministry of Science and Technology identified at least 350 sites in Iraq as being contaminated with DU. Even mild radiation is dangerous and increases the risk of cancer. In 2004, Dr Janan Hassan of the Basra Maternity and Children's Hospital said that as many as 56 per cent of all cancer patients in Iraq were now children under 5 years old, compared with just 13 per cent 15 years earlier. She stated:

It is notable that the number of babies born with defects is rising astonishingly. In 1990, there were seven cases of babies born with multiple congenital anomalies. This has gone up to as high as 224 cases in the past three years.

(Fathi, Matti, Al-Salih and Godbold, 2013: 12)

As we marked the 20th anniversary of the invasion of Iraq, we continued to see the impact of that invasion and its toxic aftermath on the security of the Iraqis, who have endured insecurity in all sectors. They have been subjected to genocide, terrorism, diseases, the killing of protesters, poverty, arrests, financial exploitation and the displacement of millions of people. When the war in Iraq officially ended in 2011 with then-US president Barack Obama declaring the withdrawal of troops, a deeply traumatised country was left behind, with a bankrupt economy. Due to falling oil prices and the effects of the COVID-19 pandemic on the country's economy, Iraq's poverty rate shot up from 20 per cent in 2018 to more than 30 per cent in 2020, meaning that 12 million Iraqis were living below the poverty line (Fordham, 2021), and the estimated youth unemployment rate was 25 per cent. The occupation of Iraq drained the country's resources and left a legacy of economic crisis, energy shortages, increased sectarianism and insurgency.

There are perhaps as many as 40 different groups (Hamourtziadou, 2014), but the major groups of armed insurgency are:

- Ba'athists, supporters of Saddam Hussein's administration, including army or intelligence officers, whose ideology is a variant of Pan-Arabism. Their goal was the restoration of the former Ba'athist government to power, and they joined forces with guerrilla organisations that opposed the US-led invasion.
- Iraqi nationalists, Iraqis who believe in Iraqi self-determination and advocate the country's territorial integrity. They also rejected the presence of the coalition forces and took arms against them.
- Sunni Islamists, Salafi/Wahhabi 'jihadists'. Salafis advocate a return to a strict understanding of Islam and oppose any non-Muslim groups and influences and regularly attack the Christian, Mandean and Yazidi communities of Iraq. They also attack Shia Muslims, whom they consider apostates. The Islamic State is the deadliest of these groups.
- Shi'a militias, including the Iran-linked Badr Organisation and the Mahdi Army. Shia Islamists are thought to be Iranian-run groups, influenced ideologically and armed by Iran.
- Foreign Islamist volunteers, including those often linked to al-Qaeda and largely driven by the Salafi/Wahhabi doctrine. They are mostly Arabs from neighbouring countries, Syria and Saudi Arabia primarily, Wahhabi fundamentalists who wish to assist the insurgency against Western forces and their allies in Iraq. They are fighting a jihad under the ideological umbrella of al-Qaeda and Ansar al-Islam.

In July 2021, as summer temperatures reached scorching levels, hundreds of Iraqis poured into the streets to protest widespread power outages in Baghdad and the country's southern provinces. In Basra, demonstrators blocked highways and burned tyres (Hamourtziadou and Gokay, 2021). There have been regular waves of anti-government protests since 2015; between September and December 2019, the country witnessed the largest and bloodiest protests since the overthrow of Saddam Hussein. Iraqis have been frustrated by the lack of clean water and electricity, widespread poverty, high levels of unemployment, government corruption and dismal prospects for the largely young population.

Since 2019, mass demonstrations have shocked areas of southern Iraq, as well as parts of Baghdad, after Iran stopped supplying electricity to Iraq because of unpaid bills. Crippled by US economic sanctions, Iran has been pressuring the Iraqi government to settle unpaid energy bills. On top of these, the pandemic left Iraq with a significantly reduced budget, preventing it from increasing both its domestic natural gas production capacity and its ability to capture associated gas and use it for power generation. A deep crisis in the Iraqi energy sector might seem to show that Iraq's energy problems can only be solved by allowing Western energy giants to run the sector, which perpetuates Iraq's dependence, lack of any real sovereignty and economic-political control by other countries and businesses.

Since the invasion, all 'democratically elected' governments in Iraq have been supported by the US–UK coalition. The new Iraq is vulnerable to internal and external threats, both military and political. Since its first democratic election, the country has shown the weakness and inadequacy of its political actors: their policies, their legitimacy, their planning and control. The country is characterised by segregation as well as inequalities of wealth and social disorganisation. The growing sectarian identities and insurgencies, among a population increasingly poor, miserable, displaced and dying in horrific daily violence, have become the biggest threat to the security of the state and society. Already impoverished under Saddam Hussein and a decade of UN economic sanctions, Iraq after the invasion lacked the capacity to cope with the total breakdown of security and it continues to suffer its effects.

The new Iraq faces threats to human rights, human dignity and human life. Its lack of development, services and resources, its food scarcity, poverty and unemployment, mean that there is still no freedom from want. The human rights violations, as well as the daily terrorist attacks, mean there is no freedom from fear. The lack of good governance in a state supported and maintained largely by occupying forces has meant that there is no freedom to live in dignity. The result is a country filled with criminals and warlords, run by a corrupt regime, with an alienated and marginalised population.

Even as the occupation was officially nearing its end, the damage done to Iraq went beyond the personal safety of its citizens; it extended to the identity and values of its communities, their sense of belonging and of being protected, their sense of living in a 'homeland'. This damage has proven to be long-lasting. In energy-rich states, oil wealth is directly linked to regime security but often fails to filter down to the wider population (Hamourtziadou, 2021). This also has a profound impact on human security, especially for those living in oil-rich regions in the south, where armed force is used to 'stabilise' the energy supply. The paradox is that Iraq, an energy-rich state, suffers from energy insecurity which affects the quality of life and hinders economic recovery.

The outcome has been countless anti-government protests, great and small, to which state security forces responded with violence.

> For a young generation that has grown up in the 16 years that have followed the toppling of Saddam Hussein, elections and representative democracy have become synonymous with corruption and MPs abusing their privileges. Religious parties, many backed by Iran, dominate the political sphere and though oil-rich Iraq has an income of hundreds of billions of dollars, the reality for many citizens is parallel with life in some of the poorest Arab nations: unemployment, a collapsing healthcare system and lack of services.
>
> (Abdul-Ahad, 2019)

Young people called for dismantling the political system. Three years later, in October 2022, protesters shouted, 'We want to overthrow the regime' (*The Guardian*, 2022).

Pro-Iran parties dominate Iraq's parliament, and more than 150,000 fighters of the former Iran-backed Hashd al-Shaabi paramilitary forces have been integrated into the state military. They were among those who killed hundreds of protesters.

The protests of September–December 2019 were the largest and bloodiest; for months, protesters took to the streets in Baghdad and towns and cities across the south of the country to demand jobs, basic services and an end to corruption. More than 500 people were killed and thousands of others wounded in clashes with security forces. 'The greatest identifiable perpetrators of violence against civilians in 2019 were government and state-associated forces, who killed some 500 protesters during May, September, October and November, including in single massacres carried out by gunmen' (Hamourtziadou, Dardagan and Sloboda, 2019). The protesters, most of them 15–25 years old, rose against lack of opportunity and deprivation, food shortages and an inefficient welfare state, all of which have left them with dismal prospects. Expressions of 'Saddam nostalgia' are even noticeable among the new generation, under 30 years old, who became young adults after the invasion.

The post-invasion war in Iraq continues to this day. Even the war's quietest months have been punctuated by moments of mass horror, and barely a day has passed without reports of civilians being shot or blown up. Despite any number of official declarations, there has been no 'turning point' towards peace, no 'mission accomplished' for 'Operation Iraqi Freedom'. An entire generation of Iraqi children has known little other than life in a country riven by violence, fear, hopelessness, internal displacement and poverty. All around them, the war's fearful legacy persists. Without any democratic process, the 'charity' or 'gift' of liberal and democratic Western states was barely disguised exploitation. In the name of that 'democratic' dream of a privatised, foreign-owned and 'reconstructed' Iraq, hundreds of thousands of Iraqi civilians have lost their lives. In a state rife with injustice, poverty, violations of human rights, government brutality and continuous foreign intervention, there can be no security and no democracy (Hamourtziadou and Gokay, 2020).,

Iraq was and still is a weak state. Between 2003 and 2023 the only constants have been the following: communal violence, terrorism, poverty, weapons proliferation, crime, political instability, social breakdown, riots, disorder and economic failure. In Iraq, we observe the lack of basic security that exists in 'zones of instability'. As in all weak states, the primary security threats facing the Iraqi population originate primarily from internal, domestic sources. In such states, the more the ruling elites try to establish effective state rule, the more they provoke insurgency. Despite it being declared a democracy, Iraq lacks regime security. In Iraq and other 'liberated and democratised' states, those internal/domestic security threats have gone hand-in-hand with the external threat posed by a collaborative external actor and the neoliberal destruction it brought to the country.

In 2021, an 'Iraq 2021 Human Rights Report' highlighted the severe and enduring injustices of the Iraqi justice system, based on 235 interviews with current or former detainees, as well as discussions with prison staff, judges, lawyers, families of the detainees and other relevant parties (United States Department of State, 2021). As reported in the Washington Post, the study detailed:

> a labyrinth of unfairness, with detainees often denied due process at every turn … Confessions frequently come through torture … [such that] detainees frequently end up signing documents admitting crimes they did not commit. Few detainees see a lawyer until they appear in court. Methods of abuse include severe beatings, some on the soles of the feet, as well as electric shocks, stress positions and suffocation. Sexual violence was also reported.
>
> (Loveluck, 2021)

In 2022, there were 1,352 arrests under the Terrorism Act (Dardagan, Hamourtziadou and Sloboda, 2023). All these men face the death penalty.

Millions of people living in war zones have been displaced by war. The post-9/11 wars have forcibly displaced at least 38 million people in and from Afghanistan, Iraq, Pakistan, Yemen, Somalia, the Philippines, Libya and Syria. The total displaced by the post-9/11 wars could be closer to 60 million, which would rival World War II displacement. 'Displacement has caused incalculable harm to individuals, families, towns, cities, regions, and entire countries physically, socially, emotionally, and economically' (Watson Institute, 'Millions Displaced by U.S. Post-9/11 Wars', 2021). At least 4 million Iraqis have become displaced since 2003, with most of them becoming refugees in neighbouring, as well as Western, countries. Refugees and asylum seekers are typically already under stress; however, they also experience 'anxiety and depression' (Strang and Quinn, 2019). While these stressors differ from other immigrants, some may be similar, such as the potential of difficulty communicating due to a different language or culture. Refugees are often seen as an 'economic burden'. This intolerance has been found in Iraqi, Syrian and Afghan refugees, who experienced 'multilevel intolerance', including but not limited to perceived prejudice, stereotypes, perceived discrimination and scapegoating (Strang and Quinn, 2019). Many vulnerable refugees face sexual abuse, robbery, resource availability and possibly death (Loescher, 2002), but these security threats are rarely discussed; more often than not, what is reported are the potential/alleged security threats generated by refugees and asylum seekers (Loescher, 2002).

Post-traumatic stress disorder

Suicide rates among active military personnel and veterans of the post-9/11 wars are reaching new peaks, according to research carried out by Ben Suitt for the Watson Institute. A report released in 2021 used governmental data, secondary literature and interviews to document a suicide epidemic that is emerging among post-9/11 fighters as part of a broader mental health crisis (Suitt, 2021). The study found that four times as many active-duty personnel and war veterans of post-9/11 conflicts had died of suicide than in combat, as an estimated 30,177 died by suicide compared with the 7,057 killed in post-9/11 war operations. It also noted that the increasing rates of suicide for both veterans and active-duty personnel were outpacing those of the general population, which was an alarming shift because suicide rates among service members have historically been lower than suicide rates among the general population. The author concludes that trauma, stress, mental health disorders, PTSD, moral injury and access to lethal means have all played a part. Moreover, the length of the war kept service members in the fight longer, providing more opportunities for traumatic exposure. 'The U.S. government's inability to address the suicide crisis is a significant cost of the U.S. post-9/11 wars, and the result is a mental health crisis among our veterans and service members with significant long-term consequences, Suitt argues' (Suitt, 2021: 28).

In the UK, the Ministry of Defence data on suicides since 1984 in the UK regular armed forces show that, since 1984, some 905 serving members of the military killed themselves while serving compared to 802 deaths related to hostile actions. Overall, the British military has suffered 802 combat deaths since 1984 in Northern Ireland, the first Gulf War One, the Balkans, Sierra Leone, Afghanistan and Iraq. This compares to 905 suicides while in service (Smith and Overton, 2022).

> In 2021, the MOD noted that in the last five years there was an increase in the rate of suicide among Army males from six per 100,000 in 2014 to 15 per 100,000 in 2018. This revelation comes as new data emerges that the percentage of British Army service personnel who have been medically discharged for mental and behavioural disorders has increased some 318% in a decade. Overall, in a decade the proportion of Tri-service discharges – Royal Navy, Army and Air Force – for mental health concerns have collectively also risen by some 245%.
> (Smith and Overton, 2022)

The Ministry of Defence released figures that showed in 2012–2013, 263 of 2,359 service personnel (11 per cent) discharged from the armed forces were for mental health issues. In 2021–2022, 38 per cent of all discharges were for such issues (505 of 1,321 discharges).

A decade ago, some 11 per cent of medical discharges in the Army were for mental health reasons, with 10 per cent and 15 per cent in the Navy and RAF respectively. In 2021–2022, this had risen to 46 per cent, 26 per cent and 43 per cent, respectively. A 2023 report by King's Centre for Military Health Research on the mental health of the UK Armed Forces found that there had been a moderate increase in PTSD in recent years in the forces, mainly among ex-service personnel. The report also notes that

> moral injury, or the psychological distress [often guilt, anger, shame or disgust] experienced after events which violate one's moral or ethical code, is experienced by UK veterans. Moral injury was found to result following acts of commission, omission and betrayal during military service. Exposure to potentially morally injurious events was significantly associated with probable PTSD, CMD and suicidal ideation.
>
> (King's Centre for Military Health Research and Academic
> Department of Military Mental Health, 2023: 6)

Jack Saul, in 'The Hidden Trauma of Moral Injury', writes of a young man who had served in Afghanistan and Iraq and years later was still haunted by the raids they had carried out.

> 'I saw how frightened the kids were as we searched the house and then arrested their father because we'd found a gun. As we took him away in handcuffs, I knew he'd probably be sent to Abu Ghraib.' His voice caught. 'Where he'd be interrogated and tortured.' He said that each time he participated in these raids – numerous times a day, day after day – he couldn't help but wonder what the U.S. was doing in this faraway land, and why he was systematically destroying families' lives. His distress deepened as he realized that he couldn't speak about any of this with any of his fellow recruits: 'staying strong' was an unspoken and unbreakable rule.
>
> (Saul, 2023)

The young man reveals that *for 10 years he's been 'wracked by guilt', he has been unable to work, and he has felt suicidal.* It is the experience of relentless guilt, shame and grief for actions, or failures to act, in wartime, which is described as moral injury, an ethical and spiritual crisis that occurs when people feel complicit in behaviours that violate their fundamental values. 'Moral injury has been recognized as one of the most significant contributors to the high rate of suicide among U.S. veterans, which is currently 17 a day' (Saul, 2023), and since the start of the War on Terror, more veterans have died from suicide than in combat.

The moral injuries of war can also be seen in the recollections of another veteran who had survived a grenade attack on his convoy, an attack that had killed his comrades. He shot and killed the boy who had thrown the grenade…

The child he shot was five years old. In the wake of this shattering experience, the veteran suffered the trauma of having come so close to being killed himself, but he was also dogged by the devastating guilt of having killed a child. Even though he could tell himself that the deaths of children were inevitable in war, when he was back in the U.S. and watched his niece and other children playing, it triggered the memory of what Iraqi kids endured due to his country having engaged in a war under false premises – and the part he played in their suffering.

(Saul, 2023)

Moral injury can occur not only when one commits violence but also as a result of *witnessing* violence and not being able to stop it. This can lead to moral and mental anguish both for combatants and for civilians. For the past 20 years, mental health in Iraq has been a difficult issue to address, as the country has faced wars, terrorist attacks, poverty and political instability. The catastrophic events that began in early 2003 have not only taken the lives of hundreds of thousands but have also affected the mental well-being of millions of Iraqis. According to Saied, Ahmed, Metwally and Aiash, in *The Lancet*, over 20 per cent of Iraqis have mental illnesses, and that percentage is steadily rising.

Fear and trauma have been widespread as the war escalated the already high rates of violence in the country (eg, bombings, shootings, and kidnappings). Millions of Iraqi people have been displaced from their homes and neighbourhoods as a direct result of the war. Increased rates of depression, anxiety, and post-traumatic stress disorder have been observed in Iraqi people who have been uprooted from their homes and lost friends and relatives.

(Saied et al, 2023: 1235)

A Baghdad-based study found that a shocking 91 per cent of women in Baghdad have experienced war-related trauma in the past two decades (Lafta and Merza, 2021). Anxiety symptoms were reported by 39.7 per cent of the women, depressive symptoms by 34.2 per cent and suicidal thoughts/attempts by 35.8 per cent. Data show that young women are more likely to attempt suicide or to have suicidal thoughts.

Many factors have played a role in persisting mental illnesses in Iraq. The invasion and occupation of the country, the unending battle with terrorist groups, the growth of sectarianism and the persistence of political instability have all contributed to a traumatising environment. In addition, the mental health system in Iraq is not well developed, and there is very little access to care and skilled personnel. Iraq passed legislation governing mental health in 2005, but it has not been fully used. 'There are only six specialised psychiatric hospitals in Iraq (two in Baghdad and four in the Kurdistan region), which do not even meet the bare minimum of the demand. Iraq has approximately 0.34 psychiatrists per 100,000 population' (Saied et al, 2023: 1236). There is a severe scarcity of mental health experts, as well as inadequate training, restricted access to resources and social stigma against working in mental health.

As the War on Terror conflicts changed, we saw the effects of drone warfare on populations. After winning the 2008 elections, President Obama embraced and increased the targeted assassination programme. Under Obama, according to intelligence analyst Robert Gates who served as Secretary of Defense from 2006 to 2011, factories were working day and night to produce the weapons to fight terrorists. 'From now', he declared, 'the watchword is: drones, baby, drones!' as he accepted an award for exemplary service to the nation and the CIA, at the Richard M. Helms Award dinner on 30 March 2011. Just days after Obama's inauguration, in January 2009, two drone strikes he authorised killed up to 25 people, including as many as 20 civilians in Waziristan, Pakistan. Neither strike hits its HVT; the first strike hit the home of 18-year-old student Faheem Qureshi, killing his cousins and friends and blinding him in one eye, while the second took the life of local elder Malik Gukistan Khan, who was a member of a peace committee. Four members of his family, his nephew and three sons (the youngest aged just 3 years old) died with him (Ahmad, 2015). The strikes continued, then doubled and redoubled, until life in Waziristan became 'hell on earth' (Rohde, 2012). The remote killing campaigns devastated the life of the society and traumatised communities as if they had been subjected to World War II-style bombings, but in ways that would be invisible to distant spectators peering at their Predator feeds. The double-tap tactic of reserving a second missile for anyone coming to rescue the wounded or retrieve the dead bodies 'put a crimp on the generosity of ordinary citizens, not to mention the Red Cross, which ordered its people to stay away from a house or car hit by drones for at least *six hours*' (Cockburn, 2015: 227). In Iraq, Airwars documented over 14,000 civilian deaths from drone strikes between 2014 and 2018 in the fight against the Islamic State. The drone strikes have caused enormous suffering to civilians: innocent people burnt alive, parents who saw their children killed, families plunged into poverty after losing their breadwinners and traumatised communities that continue to live under the threat of killer drones.

International law, casualty recording and human rights

The legal recognition of the human cost of war can be seen in the Geneva Conventions, in the emergence of the concept of war crime. The concept developed at the end of the 19th century and the beginning of the 20th century, when international humanitarian law, also known as the law of armed conflict, was codified. The Geneva Convention of 1864 and subsequent Geneva Conventions, notably the four 1949 Geneva Conventions and the two 1977 Additional Protocols, focused on the protection of persons not or no longer taking part in hostilities. War crimes are grave breaches of the Geneva Conventions of 12 August 1949, namely, any of the following acts against persons or property protected under the provisions of the relevant Geneva Convention:

- Wilful killing.
- Torture or inhuman treatment, including biological experiments.
- Wilfully causing great suffering or serious injury to body or health.

- Extensive destruction and appropriation of property, not justified by military necessity and carried out unlawfully and wantonly.
- Compelling a prisoner of war or other protected people to serve in the forces of a hostile Power.
- Wilfully depriving a prisoner of war or other protected person of the rights of fair and regular trial.
- Unlawful deportation or transfer or unlawful confinement.
- Taking of hostages.

War crimes are those violations of international humanitarian law that incur individual criminal responsibility under international law. They could be divided into (a) war crimes against persons requiring protection, (b) war crimes against those providing humanitarian assistance and peacekeeping operations, (c) war crimes against property and other rights, (d) prohibited methods of warfare and (e) prohibited means of warfare.

Prohibited acts include:

murder; mutilation, cruel treatment and torture; taking of hostages; intentionally directing attacks against the civilian population; intentionally directing attacks against buildings dedicated to religion, education, art, science or charitable purposes, historical monuments or hospitals; pillaging; rape, sexual slavery, forced pregnancy or any other form of sexual violence; conscripting or enlisting children under the age of 15 years into armed forces or groups or using them to participate actively in hostilities.

(United Nations)

As a result, several trials have taken place in the 20th and 21st centuries, where individuals were found guilty of committing some of those prohibited acts. Recently, in 2023, an independent inquiry relating to Afghanistan focused on investigating alleged extrajudicial killings by British Special Forces in Afghanistan from 2010 to 2013, examining accusations of cover-up and assessing the adequacy of the five-year inquiry conducted by the Royal Military Police. On the first day of the inquiry, Lord Justice Haddon-Cave reminded everyone in the court that the UK was a founding signatory of the Geneva Conventions. The inquiry aims to scrutinise and report any unlawful activities by UK Special Forces during deliberate detention operations (DDO) in Afghanistan. In his opening statement, Lead Counsel Oliver Glasgow KC presented seven cases, examples of DDO where alleged civilians, including children, were killed, starting with 30 November 2010, when eight men were shot dead, and ending with the killing of a married couple, on 7 August 2012, and the injury of their sleeping children. Graphic photos of the bloodied corpses were shown to the court, showing the position of the bodies, as well as the position of weapons that, it was alleged, were carried by the victims. The families insist that the dead men and boys (some of them with their hands still tied) were unarmed and that the 'evidence' was fabricated in order to support claims of lawful killings as self-defence.

There was absolute silence in the courtroom, when Richard Hermer KC, representing the bereaved families, read out the names of 33 individuals killed in seven night raids. Perhaps the most shocking incident of all was that of 18 October 2012, when the following four were killed in a room: Ahmed Shah, aged 12; Mohammed Tayeb, aged 14; Naik Mohammed, aged 16; Fazel Mohammed, aged 18.

Implausible explanations were given for how the deaths occurred. Family members provided testimony via video, claiming their loved ones had been unarmed and wondering why their home was 'raided by foreigners'. They spoke of their grief, loss and long-lasting mental trauma. One family member, whose brother and sister-in-law had been killed as they slept during Operation Cestro, was flanked by his young nephews, who (toddlers at the time) had also been shot and gravely injured, next to their parents.

What has emerged is an institutional culture that enabled wrongdoing to occur and to continue. Alarming documents point to a culture of lawlessness, operational misconduct, failure of leadership, moral corruption and disrespect for human rights. Additionally, secrecy led to a sense of exceptionality and impunity among the SAS, a toxic culture that is 'anathema to the UK's armed forces' (Hamourtziadou, 2023).

With regard to Iraq, in 2014 the European Centre for Constitutional and Human Rights (ECCHR), together with Public Interest Lawyers, submitted an Article 15 communication to the prosecutor of the International Criminal Court (ICC), alleging the responsibility of UK armed forces for war crimes involving systematic detainee abuse in Iraq from 2003 to 2008. British forces were responsible for the security of four provinces in southeastern Iraq after the 2003 invasion: Basra, Missan, Muthanna and Thi-Qar. While this responsibility was handed back to the Iraqi authorities in stages from September 2004, responsibility for security in the most violent of its domains, Basra, was the British Army's until December 2007, and UK combat forces remained in the region in an advisory capacity until July 2009. Of the post-invasion deaths from May 2003 to December 2007, 193 could be directly attributed to the Coalition military, of which 124 have been identified as victims of British military action by Iraq Body Count researchers (Iraq Body Count, 2011). A preliminary investigation was opened in May 2014 that led first to a 2017 report which announced that the prosecutor had reached the conclusion that there was a reasonable basis to believe that members of the UK armed forces had committed war crimes within the jurisdiction of the ICC against persons in their custody. The ICC 'Situation in Iraq/UK Final Report' published on 9 December 2020 makes it clear:

There is a reasonable basis to believe that various forms of abuse were committed by members of UK armed forces against Iraqi civilians in detention. In particular, as set out below, there is a reasonable basis to believe that from April 2003 through September 2003 members of UK armed forces in Iraq committed the war crime of wilful killing/ murder pursuant to article 8(2)(a)(i) or article 8(2)(c)(i)), at a minimum, against seven persons in their custody. The information available provides a reasonable basis to believe that from 20 March 2003 through 28 July 2009 members of UK armed forces committed the war crime of torture and inhuman/cruel treatment (article 8(2)(a)(ii) or article 8(2) (c)(i)); and the war crime of outrages upon personal dignity (article 8(2)(b)(xxi) or article 8(2)(c)(ii)) against at least 54 persons in their custody. The information available further provides a reasonable basis to believe that members of UK armed forces committed the war crime of rape and/or other forms of sexual violence article 8(2)(b) (xxii) or article 8(2)(e)(vi), at a minimum, against the seven victims, while they were detained at Camp Breadbasket in May 2003.

(ICC, 2020: 4)

More specifically,

the article 15 communications allege: acts of torture and other forms of ill-treatment against at least 1071 Iraqi detainees; 319 unlawful killings (267 in military operations and 52 against persons in UK custody); and rape and/or other forms of sexual violence against 21 male detainees in 24 instances.

(ICC, 2020: 11)

Crimes committed by the British included forced exertion, wilfully causing great suffering, forced nakedness and cultural and religious humiliation. This mistreatment was systematic and those who bear the greatest responsibility for the crimes are situated at the highest levels, 'including all the way up the chain of command of the UK Army, and implicating former Secretaries of State for Defence and Ministers for the Armed Forces Personnel' (ICC, 2020: 12).

Insight, and analysis, into these critical human aspects of armed conflicts and situations of violence, is provided by casualty recorders. The 53rd session of the UN Human Rights Council, in the summer of 2023, recognised the importance of casualty recording for the protection and promotion of human rights. The report of the High Commissioner for Human Rights is clear and detailed on the ways casualty recording impacts civilian protection, compliance with international law, prevention, accountability, access to services and reparations. Through its multiplicity of contexts, actors and approaches, the documenting of civilian harm can become an integral part of responses to conflict, including harm mitigation efforts, identification of remains, as well as respectful handling according to customs. The analysis and information collected by casualty recorders can guide efforts to protect civilians and ensure the enjoyment of their rights and prevent or address violations of international humanitarian and human rights law. Information collected on individual cases, as well as aggregate analysis of casualties, can indicate the severity and scale of casualties.

The European Asylum Support Office used Iraq Body Count's casualty records to assess the security situation in Iraq in determining international protection status. The independent International Commission of Inquiry on the protests in the Occupied Palestinian Territory used casualty data from several sources, including the United Nations, to investigate the use of force against protestors in Gaza. In Yemen, OHCHR casualty records revealed that children accounted for three-quarters of all civilian casualties from mines and unexploded ordnance and identified the locations of incidents. The findings were used to remind the parties of their obligations under international law to record, mark and clear landmines. In Afghanistan, Action on Armed Violence compiled and analysed casualty records produced by UNAMA, which indicated that, between 2016 and 2020, 37 per cent of civilian fatalities from air strikes were children.

> Casualty recording assists decision makers, human rights defenders, humanitarian actors and others to devise appropriate responses to civilian harm during armed conflict and threats to people during violence. Its disaggregated information, for example by age, gender, weapon type or geographic location, facilitates multiple types and levels of analysis. Casualty recording is effective at focusing the attention of the international community on the human cost of violence.
>
> (United Nations Human rights Council, 2023: 4)

A key attribute of casualty recording is that it provides victim-centred information, which is crucial for human rights and for understanding the human cost of any conflict.

Contextualising death and human suffering

Each of the 210,166 civilians whose deaths Iraq Body Count documented from March 2003 to March 2023 had a name. Each was an individual and a family member. Each had a role in the community, and each was a painful loss. Iraq Body Count researchers contextualise the dead, giving an understanding of casualties within the many contexts in which civilians find themselves in times of peace and in times of war.

Whenever possible, to respect the dignity of every human being who has suffered the biggest impact of war, and death, the dead must be named, like 3-year-old Amina Ahmed.

Each person had a role in society: a profession or occupation. Many were shepherds, politicians, shop owners and journalists. By far, the most targeted group is police officers, like Abaa Abbas Aifan and Walid Khalid Hameed: over 14,500 Iraqi policemen's deaths have been recorded by IBC. His and another policeman's death are recorded in IBC incident d12838.

The dead were members of families: mothers, daughters, fathers and sons, like the son of Zaidan Ali Wadi al-Marsoun. The death of Zaidan Ali Wadi al-Marsoum's

Table 2.12 Amina Ayman Ahmed

Age	3
Sex	**Female**
Nationality	**Iraqi**

Source: Little Amina's death is recorded in IBC incident a6262.

Table 2.13 Iraq Body Count incident a6262

Incident	**a6262**
Type	**Air attacks**
Deaths recorded	**14–16**
Targeted or hit	**Al-Aklat flour mill and house hit, casualties include women and children**
Place	**Bawabat al-Sham, Hay Al-Matahin, west Mosul**
Date	**14 February 2017**

Individuals for whom personal or identifying details were reported

IBC page	Identifying details (number if more than one)	Age	Sex
a6262-dx3523	Ali Khadr Thanon	47	Male
a6262-ec3650	Aisha Abdel Thanon	43	Female
a6262-dw3465	Hussein Ali Khadr	16	Male
a6262-zx3511	Afrah Ali Khadr	27	Female
a6262-sf3460	Noor Mohammad Hamid	20	Unrecorded
a6262-zw3445	Dalal Ayman Ahmed	5	Female
a6262-sb3646	Amina Ayman Ahmed	3	Female

Source: Iraq Body Count database.

Table 2.14 Iraq Body Count incident d12838

Incident	**d12838**
Type	**Booby-trapped house**
Deaths recorded	**2**
Targeted or hit	**Local policemen attempting to enter a house**
Place	**Albu Shijil, Saqlawiyah, north of Fallujah**
Date	**11 November 2016**

Individuals for whom personal or identifying details were reported

IBC page	Identifying details (number if more than one)	Age	Sex
d12838-kf3686	Walid Khalid Hameed	Adult	Male
d12838-fh3551	Alaa Abbas Aifan	Adult	Male

Source: Iraq body Count database.

son is recorded in IBC incident d12801, an incident where three people were killed in an air strike.

Careful, detailed casualty recording can highlight the impact of war on the most vulnerable members of society: the children. In Table 2.16, we see the recorded deaths of up to 22 children in Baghdad from infection of the intestinal tract after invading forces hit a water purification system through an air strike. The children died in hospital.

Table 2.15 Iraq Body Count incident d12801

Incident	**d12801**
Type	**Air strike**
Deaths recorded	**3**
Targeted or hit	**Victims include Mohammed Hamid Badawi Mahlawi and the son of Zaidan Ali Wadi al-Marsoumi**
Place	**Seada village, Al Qaim area, western Anbar**
Date	**13 November 2016**

Individuals for whom personal or identifying details were reported

IBC page	Identifying details (number if more than one)	Age	Sex
d12801-bv3489	Mohammed Hamid Badawi Mahlawi	Adult	Male
d12801-hz3690	Son of Zaidan Ali Wadi al-Marsoum	Adult	Male

Source: Iraq Body Count database.

Table 2.16 Iraq Body Count incident j036-i

Incident	**j036-i**
Type	-
Deaths recorded	**8–22**
Targeted or hit	**Water purification system (resulting in diarrhoea)**
Place	**Al-Alwiyah Children's Hospital, Baghdad**
Date	**4 April 2003–9 April 2003**

Source: Iraq Body Count database.

Vulnerable groups, such as the Yazidis, have also suffered. A genocide of Yazidis by the Islamic State was carried out in the Sinjar area of northern Iraq from 2014 to 2018. Iraq Body Count has recorded thousands of Yazidi deaths in incidents presented in Table 2.17.

Victims of war crimes by states and terrorist groups have been recorded. Incident k001 is an example of victims of a war crime committed by the US–UK coalition during the invasion of Iraq.

A few months after this attack, terrorists targeting UN workers bombed the Canal hotel in Baghdad, killing 22. IBC researchers collected and recorded the details of every person killed.

Victims of specific weapons, such as armed drones, were recorded during the battle for Mosul. The drone has been called a humanitarian weapon, as well as a moral one. The hunter-killer drone, writes Anderson, is 'a major step forward in humanitarian technology' (Anderson, 2010: 12). The drone, it is argued, is

Table 2.17 Iraq Body Count incident a5497

Incident	a5497
Type	**Gunfire**
Deaths recorded	**4**
Targeted or hit	**Bodies of Yazidi family members found with signs of gunshots, the family were from Khansour village, and consisted of a father, a mother and their children**
Place	**Near Hardan and Zorafa village, Sinjar, west of Mosul**
Date	**2 October 2016**

Source: Iraq Body Count database.

Table 2.18 Iraq Body Count incident k001

Incident	k001
Type	**Air strikes, including cluster bombs**
Deaths recorded	**27**
Targeted or hit	
Place	**Baghdad**
Date	**3 April 2003**

Source: Iraq Body Count database.

humanitarian as a weapon, as a means of killing, for several reasons. First and foremost, nobody dies, except the enemy: the bad, dangerous people, who are causing harm and suffering. Taking lives without even endangering the lives of one's own people (civilians and 'combatants') is good. In effect, the drone saves 'our' lives. According to Strawser, the drone is not just morally acceptable but morally obligatory. If you want to kill morally, you have to use a drone. He argues based on the 'principle of unnecessary risk', according to which it is 'wrong to command someone to take on unnecessary potentially lethal risk' (Strawser, 2010: 344). With drones, the old military dream of surgical strikes becomes a reality, and necroethics takes the form of a doctrine of killing well. It appears that technical progress and new weapons have made it possible to achieve risk-free warfare: for 'our' soldiers and for 'their' civilians. The new technology has eliminated moral concerns and dilemmas. Drones are said to have the potential for tremendous moral improvement over the aerial bombardments of earlier eras. Yet, the suffering drones caused in Iraq was immeasurable. They killed thousands of civilians, including the 14 in the incident presented in Table 2.20.

Every human being is born free and equal in dignity and rights. Even in death, human dignity should not be compromised, and the way we respect and remember our dead is central to our humanity.

Table 2.19 Iraq Body Count incident k1099

Incident	**k1099**
Type	**Suicide truck bomb**
Deaths recorded	**22**
Targeted or hit	**UN HQ**
Place	**Canal Hotel, UN HQ in Baghdad**
Date and time	**19 August 2003, 4:30 PM**

Individuals for whom personal or identifying details were reported

IBC page	Identifying details (number if more than one)	Age	Sex
k1099-dm135	Sergio Vieira de Mello	55	Male
k1099-ur270	Reham Al-Farra	29	Female
k1099-dc197	Emaad Ahmed Salman Al-Jobory	45	Male
k1099-vu184	Raid Shaker Mustafa Al-Mahdawi	32	Male
k1099-xh131	Leen Assad Al-Quadi	32	Male
k1099-vc258	Ranilo Buenaventura	47	Male
k1099-xn201	Richard Hooper	40	Male
k1099-em188	Reza Hosseini	43	Male
k1099-sk127	Ihssan Taha Husain	26	Male
k1099-es262	Jean-Sélim Kanaan	33	Male
k1099-sr205	Christopher Klein-Beekman	32	Male
k1099-zc123	Martha Teas	47	Female
k1099-sw192	Basim Mahmood Utaiwi	40	Male
k1099-zm177	Fiona Watson	35	Female
k1099-xr250	Nadia Younes	57	Female
k1099-uw119	Saad Hermiz Abona	45	Male
k1099-dn196	Omar Kahtan Mohamed Al-Orfali	34	Male
k1099-uz181	Gillian Clark	48	Female
k1099-sa254	Arthur Helton	54	Male
k1099-vs115	Manuel Martín-Oar Fernández-Heredia	56	Male
k1099-xu200	Khidir Saleem Sahir	Adult	Male
k1099-vm185	Alya Ahmad Sousa	54	Female

Source: Iraq Body Count database.

Table 2.20 Iraq Body Count incident a6079

Incident	**a6079**
Type	**Air attacks**
Deaths recorded	**14**
Targeted or hit	**Residential areas hit, casualties were mostly women and children**
Place	**Al-Rifa'i area, northwest Mosul**
Date and time	**19 January 2017, AM**

Source: Iraq Body Count database.

Table 2.21 Palestinian civilians with the surname Husuna killed in Gaza during 7–26 October

Maha Ramez Ameen Husuna	Female	18
Abdel Rahman Muhammad Mati' Husuna	Male	17
Sarah Saleh Ali Husuna	Female	15
Abdullah Muhammad Fadl Hamed Husuna	Male	13
Lama Ahmed Mahmoud Husuna	Female	13
Malik Bilal Muhammad Husuna	Female	12
Ahmed Husam Ameen Husuna	Male	12
Muhammad Bilal Muhammad Husuna	Male	11
Ahmed Hasan Ali Husuna	Male	11
Samer Ahmed Muhammad Husuna	Male	10
Muhammad Rami Muhammad Fadl Husuna	Male	10
Nour Wissam Ameen Husuna	Female	9
Bara'a Hasan Ali Husuna	Male	9
Jenna Ahmed Muhammad Husuna	Female	8
Aya Husam Ameen Husuna	Female	8
Yazan Rami Muhammad Fadl Husuna	Male	8
Huda Muhammad Fadl Hamed Husuna	Female	8
Nada Wissam Ameen Husuna	Female	7
Ali Muhammad Ali Husuna	Male	7
Jood Hasan Ali Husuna	Female	7
Munira Ahmed Muhammad Husuna	Female	6
Samar Rami Muhammad Fadl Husuna	Female	6
Yazan Muhammad Ali Husuna	Male	5
Shareef Ashraf Shareef Husuna	Male	5
Yousef Rami Muhammad Fadl Husuna	Male	5
Abdullah Muhammad Fadl Husuna	Male	4
Sarah Bilal Muhammad Husuna	Female	3
Adam Husam Amin Husuna	Male	3
Suwar Rami Muhammad Fadl Husuna	Female	1
Sabhi Hamdan Sabhi Husuna	Male	1
Niveen Khaled Saleh Husuna	Female	1

Source: Iraq Body Count, 2023.

Identifying every victim and documenting the circumstances of their death help record not only the casualties but also the suffering and the full impact of war.

The mainstream/traditional perspective

Traditional views of security regard the international society as a realm of power, interests, the struggle for survival and necessity, not morality, humanity and justice because no overarching authority exists to protect states and to resolve disputes. According to them, international relations are not amenable to moral determination, so they resist applying morality to war. This resistance is part of a general moral scepticism that is applied to international relations in general.

The reason for the resistance is twofold. In the first place, it springs from the conviction that the reality in question is morally intractable, the dynamics of international relations and war being seen to confound most, if not all, attempts to apply an alien, moral structure to them. Secondly, and more urgently, it arises from the fear that the very attempt to impose a moral solution has tragic consequences.

(Coates, 2016: 35)

The traditional approach to security can be seen as a reaction against the tendency to apply moral norms to the international domain with little regard for the many constraints of the realities and complexities of power, or the intricate 'mechanisms whereby the international order of an inferior but nonetheless real kind is sustained' (Coates, 2016, 36). This approach recognises both the possibilities and the constraints of power within an anarchic and bellicose world, as states strive to balance forces in order to bring some semblance of order, and since the balance of power involves forces in constant flux, maintaining it requires continual adjustment to changing circumstances free of moral constraints. A pure form of this state-centred approach outright rejects the subjection of politics to ethics, instead affirming the radical autonomy of politics. Morgenthau, for example, defends 'the autonomy of the political sphere against its subversion by other modes of thought' (Morgenthau, 1973: 13); 'there are no relevant standards of thought other than political ones' (Morgenthau, 1973: 11). Carr suggests, 'relations between states are governed solely by power and that morality plays no part in them (1981: 153). Carr was right about the inter-war period, and this is true today. This is, whether we like it or not, the international norm. States act based on their self-interest, or the interests of their elites. They hide behind 'higher' ideals such as human rights, but in reality, it is their crude material interests that make them start wars.

Aside from the instrumental role that morality plays in the hands of statesmen, international relations are thought to be morally indeterminable. If they cannot be morally ascertained during peacetime, they certainly cannot be morally ascertained in wartime, as war is international relations *in extremis*. The decision to go to war is not dictated by ideals and moral sentiment but by considerations of power and interest. It would then follow that next to power and interests, considerations of justice and human rights would be non-existent. As for the conduct of war – *jus in bello* – it appears to be limitless: 'War is an act of violence pushed to its utmost bounds', Clausewitz wrote (1982: 103). Instead of moral considerations, time and time again, we see pragmatism and strategy, both when war is decided and also during its course. War is simply the means to the political ends it is made to serve. As long as it retains that instrumental character, it will remain limited only by those political ends. As for the killing of civilians along the way, it is regarded as 'collateral damage', a necessary and inevitable by-product of war. It is the survival of states and that of various elites that profit from war that seems to be paramount.

The political will always be characterised by power, the struggle to obtain it and to influence the way in which it is employed (Weber, 1994), and politics always requires forcing some to act according to the will of others. However, the impact of the War on Terror on Iraq reveals more than the effects of the exercise of power; it reveals domination and coercion accompanied by lack of political legitimacy and justice. What we see in Iraq (and elsewhere) is abuse of power with devastating impact in most areas of security. What was promised in Iraq by the invading and occupying powers was freedom from a dictator's oppression: freedom from fear and freedom from want. What was instead delivered was much of the same.

Towards a human security approach

At the time of writing, it is estimated that over 1,300 Israelis, civilians and soldiers, were killed in their homes, communities and in confronting Hamas terrorists in Israel. The officially confirmed names of Israel's dead in the atrocities of 7 October 2023, and the subsequent Israel-Hamas war, were published on 19 October and continue to be updated. They include brother and sister Eitan (5) and Alin (8) Kapshitter, both killed in Dimona with other members of their family (Haaretz, 2023). On 26 October, the Health Ministry in Gaza released a document listing the names, ages and ID numbers of 6,747 Palestinian civilians (2,664 of them children) killed by Israeli forces. Table 2.21 presents one page from the 212-page document (translated by Iraq Body Count), where those killed may well have been members of the same family.

What any moral and human understanding of war needs is less concern with power and more with humility and humanity. Morality and humanity require that security is understood not at the level of states and their armies, not at the level of militant groups and their 'right fighters', but at the level of the ordinary person and their needs.

Seventy-five years ago, the UN General Assembly adopted the Universal Declaration of Human Rights, promoting international cooperation and asserting our common humanity. Yet wars continue and human lives are violated daily through military and political atrocities. This chapter began by exploring the casualties resulting from the 2001 and 2003 invasions and occupations of Afghanistan and Iraq, querying the morality and legality of the War on Terror, as well as its impact. The chapter accessed research carried out by several casualty-recording bodies and used that research to evaluate categories of human (in)security: physical, political, community and economic. It also assessed the psychological impact of war on survivors, combatants and non-combatants. After examining the relationship between international law, casualty recording and human rights, the chapter ended with a deliberation on contextualising death and human suffering in war. The human costs of the War on Terror are the physical, political, legal, mental and moral violations of human rights, freedoms and dignity: for soldiers, but – mainly – for civilians. In the absence of any human values on the side of the authorities that make decisions of war and peace, non-governmental groups such as the UN, Action on Armed Violence, Every Casualty Counts and Iraq Body Count are acting with real moral authority to demonstrate the human cost of these wars.

3 The new Cold War

Putin's war in Ukraine, 2014–2022

'The Devil's Decade' was the name given to the 1930s because of a large number of violent conflicts within and between states around the world that finally resulted in the breakout of the World War II. Now there is a general observation that the 2020s may resemble the 'Devil's Decade' more than any other period since. The emerging dynamic of global politics, sometimes referred to as 'The New Cold War', exhibits striking similarities, not necessarily to the period of the Cold War but more to the political, military and ideological battles of the 1930s, the 'Devil's Decade'.

We may be in the Second Cold War, in the sense that opposite ranks are emerging, with the USA and many Western countries on one side, and China, Russia and several other Global South powers on the other. But this is a very different sort of Cold War in which new and more dangerous and explicitly more explosive battle lines are being drawn that could last for generations. Where the Cold War was a bipolar confrontation between the USA and the Soviet Union, a competition of two clearly defined superpowers, the new Cold War has, so far, a more diverse, and to some extent, unpredictable group of actors. China's economic rise to a global superpower status and Russia's economic recovery and political determination under Putin have given rise to a multi-polar power struggle in which the USA is confronted by multiple challengers.

We are not far advanced into the 2020s, but the events and potential risks are already shaping up to be a second 'Devil's Decade', as violent hot wars have broken out in Ukraine, the Middle East and Africa, in addition to a range of bloody civil conflicts tearing apart many countries since the early 2010s. As in the 1930s, open military violence is once again becoming the usual method of dealing with international disagreements.

24 February 2022 marked the beginning of a disastrous war and the reappearance of violent armed conflict in Europe for the first time since the wars of the former Yugoslavia in the early 1990s. The Russian war against Ukraine and the resulting Western response, sharp economic embargo and open military support to the Ukrainian side, mark a turning point in international relations with far-reaching consequences for several countries and multi-level crises in numerous regions of the world. It is conceivable that we see the rebirth of a new era of conflict, the

DOI: 10.4324/9781003414827-4

end of the late 20th-century unipolar international security architecture under the hegemony of the USA, the end of globalisation that shaped the second half of the 20th century and the emergence of a new Cold War between the West and the East.

In the logic of the Cold War era, 1947–1991, conflicts do not necessarily mean war between the large powers, but rather geopolitical tension, hostility or proxy war, in which one or both of the larger powers challenges the other indirectly, as happened in Korea, Vietnam or Afghanistan. Based on the *Thucydides trap* hypothesis, Dalio (2023) predicted a higher probability of wars with economic, trade and geopolitical dimensions. The *Thucydides trap* refers to the increasing risk of a war between two sides when a rising power challenges a ruling power in a joint area of interest, influence and power. The deadly trap was first described by the ancient Greek historian Thucydides, who wrote about the conflict between two ancient Greek city-states: 'It was the rise of Athens and the fear that this instilled in Sparta that made war inevitable' (Allison, 2015).

According to Allison, the rise of China (rising contender) can be seen as such a serious threat to the USA (current hegemon) that they are on a collision course for war 'unless both parties take difficult and painful actions to avert it'. In this line, NATO increasingly warns about the threat by China, e.g. in the NATO 2022 Strategic Concept, which was immediately criticised by China (NATO, 2023). Many Western observers agree that the US–China competition will be the key determinant of geopolitics for decades to come, and it remains unclear whether Beijing and Washington will be able to escape the *Thucydides Trap*.

However, the Chinese leaders' desire to compete mainly at an economic level does not necessarily make China a geopolitical rival to the US hegemony. As witnessed over the last 20 years, Russia's resurgence has created serious concerns for the US interests in Eurasia. US Army Chief-of-Staff Mark Milley said in 2015 that 'in terms of capability, Russia is the only country on earth that has the capability to destroy the United States of America'. Russia's weapons capability and its ability to potentially wipe out the USA, as mentioned by Milley in 2015, means that in terms of geostrategic rivalry, Russia is more of a threat to the USA than China. Mark Milley further said:

> It's an existential threat by definition because of their nuclear capabilities. Other countries have nuclear weapons, but none as many as Russia and none have the capability to literally destroy the United States. ... we want over-match... we do not want a level playing field or a fair fight; we want it all in our favour.
>
> (Leipold, 2015)

Stephen F. Cohen, Professor of Russian Studies and History at New York University, has argued that the new Cold War, the Cold War of the 21st century, is more dangerous than its 44-year predecessor. He mentions specifically NATO's ever-growing military build-up in the form of more troops, weapons, warplanes, ships and missile-defence installations very close to Russia's borders as a key factor for this dangerous situation. NATO expansion eastwards was for Moscow a broken American promise. No matter what former US officials now say, Gorbachev was told by Bush and Baker in 1990–9091 that if he agreed to a reunified Germany in NATO, the alliance would not move, in Baker's words, 'one inch to the east'. When Clinton expanded NATO eastwards, for Russia he had broken a solemn promise involving its national security. Cohen further elaborated that NATO now characterises this vast Eastern front as 'its territory'. No such foreign military power has appeared so close to Russia – and to its second city, St. Petersburg – since the Nazi German invasion in 1941. 'Imagine Washington's reaction if pro-Russian bases and governments suddenly began appearing in America's sphere, from Latin America and Mexico to Canada. Of course, there has been no such discussion in the United States'. Professor Cohen further contended that the USA and NATO had disregarded Russia's concerns for years and that Putin's actions were fully understandable (Cohen, 2010; Cohen, 2017).

Many other prominent writers agreed with Professor Cohen. George Kennan, who had been the author of America's policy of containment against the Soviet Union, called NATO expansion 'a strategic blunder of potentially epic proportions'. Thomas Friedman, one of America's most prominent foreign policy columnists, declared it the 'most ill-conceived project of the post-Cold War era'. Daniel Patrick Moynihan, US politician and diplomat, warned, 'We have no idea what we're getting into'. John Lewis Gaddis, America's pre-eminent scholar of the Cold War, noted that 'historians – normally so contentious – are in uncharacteristic agreement: with remarkably few exceptions, they see NATO enlargement as ill-conceived, ill-timed, and above all ill-suited to the realities of the post-Cold War world' (Aleem, 2022).

Indeed, Western policies since the end of the Cold War can be seen as attempts by the USA and its NATO allies to establish a zone around Russia's Western borders, creating security concerns for Russia. The USA and NATO have progressively advanced the placement of their military forces and weapons closer and closer towards Russia, all the way to its Western borders (Estonia, Latvia and Lithuania were accepted in NATO in 2004, in the process of NATO's second post-Cold War enlargement). If Russia had taken similar actions with respect to US territory, for instance by placing its military forces in Canada or Mexico, 'Washington would have gone to war and justified that war as a defensive response to the military encroachment of a foreign power' (Abelow, 2022: 1).

The merger of democracy and geopolitics has had an effect that looks familiar. 'To the extent that Russia turned away from liberal democracy while Europe embraced it, it was inevitable that there would be some border between democratic and non-democratic Europe… It has been Ukraine's bad luck to have the conflict played out on its territory' (D'Anieri, 2019: 4).

Russia's war in Ukraine

The world has been plunged into a catastrophic and dangerous war in 2022. On top of the damage and the panic of the COVID-19 pandemic, the Putin regime's violent invasion of Ukraine has brought with it carnage and human suffering not experienced in Europe since the wars in former Yugoslavia. It seems that this conflict reflects the characteristics of a new era that the global inter-state system has now entered. While the COVID-19 pandemic acted as a great accelerator of all of the underlying contradictions of the system in terms of economy and human security, this war has crystallised some of them, in particular the sharpening of geopolitical tensions between the (old and new) superpowers. The Russian invasion of Ukraine in a sense has emphasised the relative decline of the US power and the absence of a unipolar world, in which one imperial power was often able to impose its will internationally as was the case during the Cold War. Major changes have been happening in the world system, namely a shift in its hegemonic structures – a shift away from North America and Western Europe and towards the emerging economies, primarily in Asia, but also towards Latin America and Africa. The eruption of the war in Ukraine is the product of this era of fast international changes. This war is historic, less because of the war itself than because of what that war is revealing about the world, and its characteristics, we live in, and about the new organisation of the world that is coming into being with new and unpredictable dangers. The USA has the essential goal of continuing as the dominant world power without being challenged by any new contender. China is working towards its goal of shaping a new global system with multiple power blocs, such as the BRICS+ bloc (Brazil, Russia, India, China, South Africa, Iran, the United Arab Emirates, Saudi Arabia, Egypt and Ethiopia). The war in Ukraine has exposed how far along we are in this global shift, in this military, political and economic competition.

Putin undertook the invasion, driven by Great Russian chauvinism and a desire to establish an expanded Russian sphere of influence, a zone of friendly states between Russia and Western powers on Russia's border. He has thus revived the ideology of 'Novorossiya' – 'New Russia' – with the intention of establishing an expanded area of Russian language, culture and the assimilation of states or statelets into such an amalgam, a security and cultural zone, an expanded sphere of influence. This aggressive Russian policy (and practice) follows three decades of provocative eastwards expansion by NATO (the US-led Western military alliance), as expressed clearly by many prominent Western observers too.

> Over the last 30 years, multiple Russian leaders have warned about the threat NATO expansion poses to Russian security. In 1995, Russian President Boris Yeltsin informed U.S. President Bill Clinton that 'the borders of NATO expanding towards those of Russia' would 'constitute a betrayal' and 'humiliation' of the Russian people; in 1997, former Soviet leader Mikhail Gorbachev warned the U.S. Congress that he believed NATO expansion was 'a mistake, it is a bad mistake, and I am not persuaded by the assurances I hear that Russia has nothing to worry about'; and in a famous speech at the Munich Security Conference in 2007, Russian President Vladimir Putin labelled NATO expansion a 'serious provocation that reduces mutual trust.' Indeed, in December last year, just two months before it invaded Ukraine, Russia publicly published proposed draft treaties with NATO and the US, demanding an end to any further

eastward expansion of the alliance – a demand which NATO unanimously rebuffed, and which the U.S. in particular rejected as a 'complete non-starter'.

(Moller-Nielsen, 2022)

Many other countries have close economic and political relations with both sides, and some face sanctions and retaliation for whatever choices they make. Around this war, there may be an expanded historical alliance with the US position globally, but there is also growing concern about America's changing global position, its relative decline and emerging isolationism since Trump's presidency. Even among the close allies of the USA, there is an irritation that US efforts, and open hostility towards Russia and China, in order to secure mainly its state interests, are hurting many of its European allies. In this context, the 2022 war in Ukraine has exposed economic nationalism, a 21st-century version of 17th-century mercantilism, as the new ideological reality affecting many states and dominating the inter-state world system as more and more nations move to use national policies to promote exports while creating barriers to imports. The war in Ukraine, and responses to it, seems to have exposed the difficulties the USA is now being faced when trying to convince its allies to support its policies. All these remind us that for a declining British Empire it took two wars, 1776 and 1812 in North America, before it realised its changed (declined) position in the new world order, after initial feelings of disbelief, treachery, outright fury and bitterness. 'Although it took over 400 years to build the British Empire to its peak, it only took a few decades until it started to unwind' (Swain, 2008).

The way the war affects Ukrainian people has been truly appalling, but this is not the first time we have witnessed such tragic and criminal behaviour in our world, even since the beginning of this century. Baghdad, Kabul, Fallujah, Mosul, Aleppo, Sirte and Tripoli were all bombed and shelled into rubble with hundreds of thousands of civilian casualties. The wider reality of the terrible suffering in Ukraine is that it is not an unprecedented brutal war, but that it is all too typical of many other wars we witnessed in recent years. All the atrocities seen in Ukraine have been carried out in other wars, extending to the deaths of hundreds of thousands, even millions of people.

Two years after Putin's forces rolled over through Ukraine, the war is beginning to look more and more like Syria. The chances of either Russia or Ukraine winning a decisive victory have already disappeared. To ensure a decisive military victory on the battlefield in the near future sounds like wishful thinking. Those who call for total victory over Russia are being as unrealistic as Putin was in February 2022 when he tried to control the whole of Ukraine with fewer than 200,000 soldiers. More likely the war in Ukraine is a grinding conflict with no outright winner, not so different from the long open-ended wars that lasted decades in the Middle East.

Many people think that the war in Ukraine started with the Russian invasion in February 2022. Secretary General of NATO, Jens Stoltenberg, disagrees. He said that 'the war didn't start on the 24th of February last year [2022]. This war actually started in 2014, with the illegal annexation of Crimea and then a few months later, Russia went in and took control over eastern Donbas'. Yes, the fighting on

Ukrainian soil started on 12 April 2014, when a 50-man commando unit headed by a Russian army veteran, Igor Girkin, seized Sloviansk in Donets oblast, and since then the last nine-plus years, NATO allies have been providing support to Ukraine with equipment, money and training (NATO, 2023).

Indeed, the war started in eastern Ukraine in April 2014 with low-level fighting between the Ukrainian army and Moscow-backed separatist rebels from the region who soon after seized some more towns in predominantly Russian-speaking eastern Ukraine. It escalated to an outright war between Moscow-backed separatists and the Ukrainian army. More separatist rebels began popping up in eastern Ukraine shortly after Russia had invaded and annexed Crimea in March 2014. They seized towns like Sloviansk and Donetsk, in the eastern region of Donbas, in outrage against the protests that had toppled Ukraine's pro-Russia President Viktor Yanukovych, himself from that same region. The way that Russia seized Crimea by force from Ukraine in March 2014 was aggressive and illegal; there is no doubt about this. But the actual question of whether Crimea is deep-down Russian or Ukrainian is much less clear. Crimea has technically been part of Ukraine since 1954, when Soviet leader Nikita Khrushchev transferred it from the Russian Soviet Socialist Republic to the Ukrainian Soviet Socialist Republic, both being part of the Soviet Union. The reasons for this were more related to internal Soviet calculations rather than whether the Crimean population were Ukrainians or Russians. Also, it did not matter much then, since both 'republics' were part of the same state, the Soviet Union. In 1991, when the Soviet Union broke up, the general expectation was that Moscow to demand Crimea back. This was because the majority of Crimeans were/ are Russian, not Ukrainian, and also the strategic location of Crimea was essential for the Russian navy. Historically, too, while Crimea has been changing hands between regional powers for centuries, for most of the last 200-plus years it has been part of Russia. That is why everyone expected Russia to take it back when the Soviet Union broke up in 1991. However, Russia was then in the middle of a serious economic crisis with a weak government and huge political uncertainty; therefore, Moscow had more urgent priorities and did not make that demand in 1991.

> Even though Putin's annexation of Crimea was, by most accounts, illegal under the United Nations Charter and customary international legal norms, his actions should nevertheless come as no surprise to the West. Putin has been decrying the West's eastward movements for a while now, and he is not without justification in his fear of being geographically surrounded by international powers greater than Russia.
>
> (Cook, 2015)

Whole scale Russian military invasion started eight years after the annexation of Crimea and the fighting in eastern provinces erupted in 2014. During those eight years up to February 2022, the conflict in eastern Ukraine had already killed over 14,000 people. Particularly, fierce battles in 2014–2015 ended with one-third of the region's territory occupied by two Russian proxy statelets, the Donetsk and

Luhansk People's Republics. During the months between September 2014 and February 2015, Russia, Ukraine, France and Germany signed several agreements, which came to be known as the Minsk Agreements after the Belarusian capital where it was settled, as a result of which forward movement of troops was stopped and fighting was reduced temporarily[1]

In the first Minsk (Minsk-1) negotiations, Ukraine and the Moscow-backed separatists agreed a 12-point ceasefire deal in the Belarusian capital Minsk in September 2014. The provisions of this deal included prisoner exchanges, deliveries of humanitarian aid and the withdrawal of heavy weapons. This ceasefire agreement came five months into a conflict that had by then killed more than 2,600 people.

The Minsk-1 deal consisted of 12 short general points, which addressed:

- Practical aspects of implementation of a ceasefire.
- Impunity for offenses committed during military operations in the self-proclaimed republics.
- Transfer of power from central to the regional level (decentralisation) in Ukraine.
- Determination of the interim status of the self-declared republics.
- Organisation of local elections in self-proclaimed republics.
- Improvement of the economic and humanitarian situation in Donbas.

However, the agreement broke down soon after, with violations by both sides. Minsk-2 came few months after, in February 2015, where representatives of Russia, Ukraine, the Organisation for Security and Cooperation in Europe (OSCE) and the leaders of two pro-Russian separatist regions met and signed a 13-point agreement (Reuters, 2022). Minsk-2's 13 points were:

- Immediate, comprehensive ceasefire.
- Withdrawal of heavy weapons by both sides.
- OSCE monitoring.
- Dialogue on interim self-government for Donetsk and Luhansk, in accordance with Ukrainian law, and acknowledgement of special status by parliament.
- Pardon, amnesty for fighters.
- Exchange of hostages, prisoners.
- Humanitarian assistance.
- Resumption of socioeconomic ties, including pensions.
- Ukraine to restore control of state border.
- Withdrawal of foreign armed formations, military equipment, mercenaries.
- Constitutional reform in Ukraine, including decentralisation, with specific mention of Donetsk and Luhansk.
- Elections in Donetsk and Luhansk.
- Intensify Trilateral Contact Group's work, including representatives of Russia, Ukraine and OSCE.

The leaders of France, Germany, Russia and Ukraine gathered in Minsk at the same time and issued a declaration of support for the deal. Many military and political steps of the deal, however, remained unimplemented. A major obstacle was related to the fact that Russia kept insisting that it was not a party to the conflict and therefore not bound by its terms. Both antagonists, Moscow and Kyiv, interpreted the agreement very differently, which led to what was dubbed by observers as the 'Minsk conundrum'. Ukrainian government saw the Minsk-2 agreement of 2015 as an instrument channel to re-establish its control over the rebel-controlled territories in the east. It was in favour of a ceasefire mainly to establish control of the Russia–Ukraine border and was willing to give only a very limited devolution of power to the regions controlled by pro-Russian separatists. The Russian government, on the other hand, viewed the deal as obliging Ukraine to grant pro-Russian rebel authorities in Donbas extensive autonomy and full representation in the central Ukrainian government (which would lead to reducing the power of central Ukrainian authorities), and only after these steps would Russia return the Russia–Ukraine border to Kyiv's control (Al Jazeera, 2022).

Western responses to these attempts, the Minsk agreements of 2014 and 2015, were in simple terms inconsistent and imprecise. Duncan Allen of the Chatham House summarises this position:

> US and EU policies have not been as clear and as consistent as they could and should have been or as policymakers like to think. Some decision-makers focus on making Russia change its policy (and, failing that, punishing it with economic sanctions); some would be content if Ukraine gave in to Russia's agenda; some want a compromise (which they rarely define); some face two ways, voicing support for Ukraine's sovereignty while hoping that an agreement acceptable to Russia is possible; and others seem prepared to accept anything that returns Europe to what they consider normalcy. In so far as there is a prevalent view in Western capitals, particularly in Europe, it is that implementation means identifying a point between the Russian and Ukrainian positions as regards elections and special status.
>
> (Allan, 2020)

While Western policymakers were trying to find a way to implement Minsk-2 so that everyone would get something and (as a result) the actual fighting would stop, the Russian side kept things clear and consistent and maintained that Ukraine must agree to the Russian interpretation of the Minsk-2 before Russia could even notionally clear the border.

As a whole, these agreements were never implemented, and soon after the fighting in the form of a trench war continued. Roughly 75,000 troops facing off along a 420-km-long front line cutting through densely populated areas. During this eight-year of fighting in eastern Ukraine, the region's economy and heavy industries were seriously damaged, and millions of people were forced to relocate. The intensity of the fighting differed from month to month, but there was still a large level of casualties, many of them civilian. By the time the Russian troops actually invaded Ukraine on 24 February 2022, more than 14,000 people had been killed in the east of Ukraine (Crisis Group, 2024).

The politics of the past

To garner internal support, so as to justify Russia's aggression that had already resulted in thousands of deaths by 2022, and to justify yet another unjust war with further atrocities, the Russian leadership employed the two best weapons available to all leaders of all nations: nationalism and history. In *Memory Makers: The Politics of the Past in Putin's Russia*, McGlynn writes:

> Russian official and societal obsession with sanitizing history and moulding it into something usable to prove exclusive heroism and victimhood, is fuelled by an insecurity borne of changing ideological regimes and the senselessness of the historical traumas Russia experienced in the twentieth century. The power of these cultural memories is immense and the Kremlin has wielded this power to prepare its nation for war and repression.
>
> (2023: 2)

Memory politics is at the heart of the Russian political system. In 2020, Putin brought in new legislation that not only allowed him to stay in power until 2036 but also contained a codification of the duty to 'defend historical truth' and 'protect the memory' of the Great Patriotic War, what Russian call the 1941–1945 Soviet Union's war against the Nazis. This was the Kremlin's 'call to history' that defined what it meant to be Russian, that justified its own rule and its aims to project power both domestically and internationally. The Russian government 'pushed history into the heart of Russian political and popular culture' (McGlynn, 2023: 3). Historical narratives – Kremlin's creation of a usable past – became part of political discourse and discussion on matters of existential significance. Those historical narratives combined and distorted knowledge of Russia's history, politics, culture and national identity. McGlynn calls this 'historical framing': the process of making history, or the past, relevant to people by pretending that past traumas and triumphs were repeated and were being watched in real/present time. The invocation of Russia's history was aimed at legitimising government policy, while at the same time constructing an understanding of patriotism. Those historical narratives served as guides for Russians to help understand and interpret the world – and their country's role in it. Russian scholar Malakhov has argued that the Russian government turned to history in order to feed a public appetite for a narrative that presented Russian identity as worthy of pride (Malakhov, 2018).

Encouraging that pride, the government – supported by state media – promoted narratives that would appeal to most people, to as many ideologies and political persuasions as possible: imperialists, communists and ethnonationalists (Government.ru, 2013). What was presented and promoted through the use of history was a culturally, politically and ethnically inclusive image of Russianness, creating a national identity from which political legitimacy could be derived in the present. As Benedict Anderson argued, for nations – all nations – to exist, they must connect their dead, their living and their yet unborn (the nation's future generations) to an imaginary or imagined community (Anderson, 1983). This requires what Israeli historian Zerubavel calls a 'master commemorative narrative' that constructs and provides a notion of a shared past (Zerubavel, 2005). By using history, the pro-Kremlin master commemorative narrative supports three arguments: that Russia needs a strong state, that it has a special path of development and that it is a great power with something unique to offer the world (McGlynn, 2023).

This historical framing was clearly seen in the conflation of the Ukraine Crisis and the Great Patriotic War by the Russian government and media in 2014, when the Kremlin used the spectre of the Nazis and wartime Ukrainian nationalist Stepan Bandera (the term *Bandera* has long been used in Russia as a byword for traitor) to undermine the Maidan protests and justify the annexation of Crimea, framing the conflict 'as a rerun of Soviet-Nazi battles in Ukraine, with Russian-backed combatants cast as Red Army soldiers valiantly fighting fascism' (McGlynn, 2023: 23). On 24 August 2014, Russian forces paraded dozens of Ukrainian POWs through the streets of Donetsk. As they trundled past the crowds, some bandaged, some limping, the men were marched past the remains of Ukrainian armoured personnel carriers destroyed in the battle and put on display in the city's main Lenin Square. Hundreds of people lined the street to see the soldiers who walked with their heads bowed and their hands behind their backs, flanked by men carrying Kalashnikovs. Kremlin-backed militiamen were bedecked in St George Ribbons, the commemorative symbol of the Soviet victory over Nazism. The highly stylised parade was meant to recreate the forced march of nearly 60,000 German Nazi soldiers through the streets of Moscow in 1944. Some threw bottles at them, some waved the Russian flag and the red, black and blue standard of the self-proclaimed Donetsk People's Republic, while others shouted 'fascists'. Street cleaning vehicles moved behind them spraying water where they had walked, similar to what happened in Moscow in 1944. This comparative narrative of historical analogies peaked in the run-up to the annexation of Crimea and the referenda in the Donetsk and Luhansk People's Republics.

The second area of historical framing was the Russian coverage of Western sanctions, where Soviet nostalgia and Soviet collapse were exploited to present sanctions as a Western attempt to humiliate and damage Russia, as in the 1980s and 1990s. Putin was presented as an avenger of the degradations and harm that Russia had suffered. He emphasised US involvement through its sponsoring Islamic extremist movements to fight the Soviet Union and Russia, groups that originated in Afghanistan. He alluded to US complicity in the Chechen conflicts and claimed that the world was witnessing a return to the Cold War as a result of actions and policies taken by the USA.

Russia was depicted as a victim of Cold-War era demonization, feeding into a narrative of Russophobia. (…) He also drew attention to Russia's nuclear superpower status, mirroring Cold War rhetoric and undermining his argument that the West bore sole responsibility for reviving tensions.

(McGlynn, 2023: 78)

This was a post-revisionist interpretation of the Cold War, in which the conflict was allegedly driven primarily by Russophobia. By resisting, Russia was defending itself from Western attack and regaining prestige.

The third example is a study of how Russian media covered military intervention in Syria, depicting the intervention as 'a just restoration of the post-Yalta order and Soviet great power status during the immediate post-war period' (McGlynn, 2023: 24). The war in Syria was presented as a proxy war within the context of the Cold War policy of containment. Russia's military intervention in Syria began at the request of President Assad to help the Syrian government fight jihadists. Russian military activities were depicted as a war against ISIS, rather than for Assad (Notte, 2016). Relations with the West, which was at the same time fighting a War on Terror, were an important part of a narrative, where any of the West's acceptance of Russia's efforts and renewed power internationally meant a resumption of Cold War strategic balance, and where any resistance or criticism was interpreted as Cold War aggression and containment. The Russian media presented the intervention in Syria as evidence of Russia regaining its status and breaking the alleged containment and isolation policy announced by the West after 2014 (Babayan, 2017). The Syria coverage promoted the notion that Russia was restoring elements of Soviet greatness, while fighting terrorists, projecting its geopolitical and military might. The intervention in Syria gave 'a tangible reality to Moscow's concept of a new international order' (Pierini, 2015).

In September 2015, the Kremlin narrative cast Russia as forming or joining an international coalition against terrorism, as a legitimate and respected member of the UN. To strengthen this image of Russia, writers and speakers drew on World War II's memories: the anti-Hitler coalition, the Yalta conference and the founding of the UN, of which the Union of Soviet Socialist Republics was an important and mighty member.

> By invoking the Allies, Russian politicians were ostensibly trying to encourage the West to join an anti-ISIS coalition to defeat terrorism in Syria. Given Western hostility towards Bashar Al-Assad, the Russian government must have understood that this was unlikely to succeed, but it was a useful way to reiterate to domestic audiences that Russia had inherited a right to sit at the table of the major world powers.

(McGlynn, 2023: 92)

In a speech to the UN, to mark its 70th anniversary, Putin stressed the need for collaboration and partnership: 'similar to the anti-Hitler coalition, it could unite a broad range of parties willing to stand firm against those who, just like the Nazis, sow evil and hatred' (Putin, 2015: 4). Putin's UN speech has been referred to as

'Putin's doctrine'. Following this speech, Russia intensified and expanded its military presence in Syria. More than 45,000 air strikes were declared in just the first three years of fighting – many targeting civilian areas and infrastructure. More than 24,000 civilians have been locally allegedly killed either by Moscow's actions or in events where communities were unable to distinguish between Russian and regime attacks. Russia itself still publicly maintains no civilians have been harmed in its strikes (Airwars, 2024).

What does Putin want? And why did his armies invade Ukraine?

Many Western commentators claim that Putin's decision to invade Ukraine was part of an imperial Russian plan to expand into neighbouring lands. William Courtney of the RAND Corporation argues that,

> Russia's revanchist and imperial ambitions may not stop at Ukraine. Unless Russian forces are defeated in Ukraine or withdrawn by new Kremlin rulers, Moscow might assault other post-Soviet neighbors. The West may face limits on the extent to which it could help them thwart such attacks.
>
> (2023)

'Putin is an imperialist who must be stopped now, or he will become more dangerous', says US Senator Chuck Grassley (2023). According to Patrick Smith of NBC News, Russia's invasion of Ukraine has raised fears that Putin is 'intent not only on claiming its neighbor and former Soviet republic, but potentially has his eye on Poland, Finland and the Baltics, among others' (Smith, 2022). It is now clear that 'Putin's endgame is nothing short of a revanchist imperialist remaking of the globe to take control of the entire former Soviet space', says Evelyn Farkas, an American national security advisor (Politico, 2022). As for Strobe Talbott, former US deputy secretary of state from 1994 to 2001, he believes that 'Putin certainly has an endgame in mind: It's recreating the Russian Empire with himself as tsar' (Politico, 2022). After Ukraine, the Kremlin's next targets could be Moldova and the Baltic countries, Admiral Michel Hofman, Belgian Chief-of-Defense, warns (Hulsemann, 2023).

To Alexander J. Motil, Rutgers University political scientist, Putin's Russia is trying to capture other countries just like Hitler attempted in the 1940s. He finds striking similarities between Vladimir Putin's Russia and Adolf Hitler's Germany, which, he says, are not accidental.

> Both regimes had – the past tense is intentional – the same historical trajectory because both were the product of imperial collapse and its destabilizing aftermath on the one hand and the emergence of a strong leader promising to make the country great again on the other.
>
> (Motil, 2022).

According to Jonathan Katz, a German Marshall Fund senior fellow and director of Democracy Initiatives, Putin 'is this century's equivalent to Hitler, and the threat he poses to Europe, U.S. and global security extends far beyond the current conflict in Ukraine' (Herman, 2022). 'Putin is proving to be the 'Hitler of the 21st century' with the invasion of Ukraine, says Leo Varadkar, Ireland's Deputy Premier (Independent, 2022). Even Prince (now King) Charles of Britain, during a visit to Canada in 2014, carelessly declared 'and now Putin is doing just about the same as Hitler'. He was said to have expressed this to Marienne Ferguson, a museum volunteer who fled to Canada with her Jewish family when she was just 13 (Quinn, 2014). Many others in the West suggested that Putin was insane or dying or trying to resurrect the Tsarist Empire or the Soviet Union.

Is this how we can explain the reasons why Russia invaded Ukraine in 2022? Is that because Putin has wanted to recreate a Russian Empire by invading surrounding countries? Is he the Hitler of our times?

Putin's Russia today, after more than 20 years under his leadership, is clearly a closed and repressive regime, but there are lots of different kinds of oppressive regimes, of which Hitler's fascism is just one variety. There is a tendency among Western observers to see every conflict as a new World War II. Uncritically comparing every crisis with World War II is not only a lazy but also a dangerous habit. Most of the Western policy and media elite seem to have been bought into this simple-minded view. While Putin is a cruel, despotic and brutal leader, he is not genocidal in the way Hitler's regime was. All this frantic eagerness to explain Putin's policies through World War II experience with Hitler and Nazi Germany does not at all help us to understand what motivated Putin's regime to invade Ukraine. Comparisons with Hitler or the Nazis are an unimaginative and imprecise way of seeing the war in Ukraine.

Putin's rise to and consolidation of power resemble a rather conventional autocrat content with personal power and enrichment, but he doesn't have geopolitical goals and ideological motivations comparable to Hitler's Third Reich. Historian Richard Evans says:

Putin's aims are limited. They're very ambitious, but they have limits. Hitler's aims were unlimited. He literally wanted to conquer the world, and his central belief was the racial question – he saw history in terms of racial struggle. Putin, however, is a Russian nationalist. He believes that Ukrainians are Russian, not that they are an inferior race.

(Millan, 2023)

Similarly, for Rajiv Sikri, an Indian diplomat with extensive experience in the region, Putin's aims are limited. He believes that Putin probably wants a pro-Russian (or at least not hostile to hostile to Russia) government in Kyiv, and – at the very least – an overall neutral Ukraine.

But Putin is not going to let go of Crimea, and Luhansk and Donetsk will become puppet states, independent in name but effectively part of Russia like South Ossetia and Abkhazia. They will be the bases for further encroachments on territory in Ukraine that they claim. Putin's interest in Ukraine is limited to the eastern, Russian-speaking parts of Ukraine, not Western Ukraine which has dominated Ukrainian politics since the Maidan revolution of 2014.

(Politico, 2022)

Putin became Putin, an effective hard-line leader, by first levelling Grozny, the capital of Chechnya, a tiny Muslim republic in southern Russia with just 1.5 million people, in 1999. No other city, since the end of World War II, was so intensely bombed. Other cities and towns in Chechnya too were reduced to rubble, and thousands of Chechen fighters and tens of thousands of civilians were killed. 'Fast becoming the Russian everyman and all-action hero in the defence of the motherland, ... Putin handed out hunting knives as Christmas presents to Russian soldiers serving in Chechnya' (Bellamy, 2023: 42).

In March 2000, days before the presidential election, a triumphant Putin, then the Acting President, flew to Grozny in a Russian SU-25 fighter jet, in a full pilot suit, seated in the co-pilot's seat, to commemorate his victory over Chechen rebels. Thomas de Waal, a journalist who covered Chechnya in the 1990s, sees many similarities between Putin's war in Chechnya in 1999 and the war in Ukraine in 2022.

The use of heavy artillery, the indiscriminate attacking of an urban center. They bring back some pretty terrible memories for those of us who covered the Chechnya war of the 1990s. ...There was a project to restore Chechnya to Russian control, and nowadays in 2022, to restore Ukraine to the Russian sphere of influence... And there was no Plan B. Once the people started resisting, which came as a surprise in Chechnya and is coming as a surprise in Ukraine, there was no political Plan B about what to do with the resistance.

(in Myre, 2022)

Since he was first appointed by Yeltsin to be the Prime Minister in 1999, and Acting President later in the same year, Putin ruled the country as an authoritarian Russian nationalist. He was not ideologically motivated; he was a pragmatic leader. 'When George W. Bush announced his 'War on Terror', Putin was quick to insist that Russia was already at war with radical Islamism, in Chechnya' (Bellamy, 2023: 50). Later, in his war against Ukraine, Muslim Chechen army was one of the most loyal and reliable for Putin's campaign. 'The Chechen forces deployed to Ukraine were at the heart of some major events...' Their reputation preceded them, and they were termed 'elite' and 'battle-hardened'. [Chechen leader] 'Kadyrov's forces have provided the Russians with a group of willing, motivated and well-trained personnel to conduct difficult operations' (Cranny-Evans, 2022).

Putin's brand of authoritarian Russian nationalism has a 'pick-and-mix approach to ideology' (Faure, 2022). There are various historical figures, both from the Tsarist Russian and Soviet periods, and certain ideologues he takes or opportunistically picks some aspects in shaping his not-so-clear ideology. Probably a key driver of Moscow's policy more than one single figure, or one single ideological tradition, is one shady right-wing think tank, which may give us some ideas of Putin regime's pragmatic ideological roots – the Izborsky Club.

The Izborsky Club (*Izborskii klub*) – 'a call from the past'

Vladimir Putin's brand of authoritarianism has a pick-and-mix approach to ideology. But the greatest influence on his thinking has been a rightwing think tank, the Izborsk Club.

(Faure, 2022)

Launched at the end of 2012, the Izborsky Club stands as a symbol of the ideological roots which can provide some elements to understand Putin's decision to invade Ukraine. The club advocates Eurasianism,[2] expanding Moscow's control and influence over a region encompassing the former Soviet Union. This think tank provided a home to a large group of self-identified nationalists and anti-liberals with the central objective of influencing the direction of the Russian state. In a way, the Izborsky Club was 'the new conservative avant-garde', which merits comparison with the New American Century project, PNAC, bringing different conservative movements around a single platform (Laruelle, 2016: 630).

The club was launched, with the aim of unifying patriotic Russians, in 2012 in the small town of Izborsk[3] (Pskov oblast, north-western Russia), during the celebration of the 1,150th anniversary of the city, which gave the Club its name. Soon after, further Club meetings were held in other places, such as Yekaterinburg, Ulyanovsk, St. Petersburg, Saratov, Bryansk, Belgorod, Tula, Kaluga, Omsk, Nizhny Novgorod, Orenburg and Donetsk, as well as Yakutia, Dagestan and Crimea. The Club has significant financial resources and strong ties to the Kremlin. Even though Kremlin spokesperson Dmitry Peskov denied the existence of any links with the Izborsky Club, it received 10 million rubles worth of grant from the Presidential Administration as a non-profit organisation. Its opening meeting was attended by Vladimir Medinsky, then the Minister of Culture of the Russian Federation and now a personal advisor to Putin, as well as the governors of many regions and presidents of state republics, such as Yakutia, Dagestan and Chechnya (Zygar, 2023). Aleksandr Prokhanov, the Club's chairman, declared in his speech that

Our club is a laboratory, where the ideology of the Russian state is being developed. It is an institute where the concept of a breakthrough is created; it is a military workshop, where an ideological weapon is being forged that will be sent straight into battle.

(Calabresi, 2014)

In June 2014, a delegation of more than two dozen leading members gathered at the Livadia Palace in the Crimean Peninsula, which was once the summer retreat of Russia's tsars. The attendees ceremoniously kissed the Crimean soil and had a tour of one of the battleships of Russia's Black Sea fleet there (Faure, 2022).

The Izborsky Club, from the start, aimed to bring together hardliner socialists and Soviet patriots together with monarchists and Orthodox conservatives, and in this way to generate a reconciliation between the Reds and Whites – combining symbols of 'Russian soul' coming from the Tsarist period and the Russian Orthodox church with the Soviet military and industrial images. The launching of the Izborsky Club provided a platform to several key intellectual figures, from right-wing conservative Russian nationalists to hardliner Stalinists, together with some key Moscow power players. Alexandr Prokhanov, veteran nationalist author, campaigner and editor of the newspaper *Zavtra*, was its founder and chairman. Prokhanov has been closely connected to some influential Orthodox businessmen such as Konstantin Malofeev, who is allegedly one of the main funders of the Donbas insurgency in eastern Ukraine. Among leading members were Orthodox priest and best-selling author Bishop Tikhon, rumoured to be Putin's personal confessor; economist and politician Sergei Glaz'ev, an adviser to Putin; Moscow State University professor, right-wing philosopher and foremost Eurasian strategist Alexander Dugin, known as 'Putin's brain'; oligarchs Oleg Rozanov, Yuri Lastochkin and Aleksandr Notin; Communist Party leader Gennady Zyuganov; Kremlin's Islamic-world strategy coordinator, Uzbek Shamil Sultanov; and leading TV news anchors Mikhail Leontev and Maksim Shevchenko (Laruelle, 2016).

Putin, keeping his distance, has never attended the Club's meetings, and the Club has not been so closely identified with him as the Head of State. However, this perfectly suited his preference for pragmatism rather than being closely associated with the Club's clearly right-wing Russian imperialist grand schemes. Rightly, many observers described Putin as 'an opportunist rather than an ideologically driven strategist'. However, this does not change the fact that 'the discourse and narrative frame' of the Club both feed into and reflect the narrative of Putin's regime (Bacon, 2018). With the annexation of Crimea and the start of the conflict in Ukraine in 2014, the Izborsky Club's agenda of the unity of the 'Reds' and 'Whites' appeared to fit within the Putin regime's long-standing commitment to national unity. In 2014, in his speeches, Putin closely echoed the Club members' language and style, arguing that Crimea constituted the spiritual and political centre of Russia. The Club's 2016 *Doctrine of Russian World* document included 'the protection of ethnic Russians' rights against the 'Russophobia' of the Ukrainian ruling elites, dominated by 'neo-Nazis' (Faure, 2023). In October 2021, the Club issued a new manifesto called *Ideology of Russian Victory*, which can be considered the most elaborate doctrinal platform justifying the war in Ukraine (Laruelle, 2022).

War damage and casualties

More than two years have passed since 24 February 2022, the Kremlin's full-scale invasion of Ukraine. There are already more than half a million casualties, Ukrainian and Russian, including almost 200,000 dead, the vast majority combatants – that is the shocking toll to date of the casualties of the 2022 war in Ukraine. In the war's first year and a half, Ukraine's military casualties had already surpassed the number of American soldiers who died during the nearly two decades of the war in Vietnam (*New York Times*, 2023). Russian military personnel losses have so far outpaced Ukrainian personnel losses. The *Guardian* newspaper reported on 6 January 2024 that 'the average daily number of Russian casualties in Ukraine has risen by almost

300 during the course of 2023' (2024). Both the Ukrainian government in Kyiv and the Russian government in Moscow are downplaying the number of their soldiers killed and wounded; however, according to the report by *Le Monde*, 'casualties are comparable to those of the First World War'.

With no reliable record available, international organisations, the media and allies have been making their own calculations for the past 18 months, offering widely divergent estimates. The latest comes from US officials quoted on condition of anonymity by the *New York Times* on 21 August. They put the number of soldiers killed and wounded on both sides since February 2022 at 500,000 (Vincent, 2023).

Two years after the start of the war, no end to the carnage is in sight and calls for a negotiated solution are wishful thinking at this point. It is obvious that both sides are losing. *Le Monde* reported in July 2023 that a French general, Jacques Langlade de Montgros, warned that the conflict in Ukraine 'is a war of attrition, set for the long term like two boxers in a ring, exhausting each other blow by blow, not knowing which one will call first' (Lynch, 2023).

The IMF forecasted a 35 per cent contraction of Ukraine's economy in 2022 as a result of the war. 'With the war ongoing, the situation remains extremely fluid, and any forecast is at this stage subject to massive uncertainty', the report said, predicting the economy could contract by between 25 per cent and 35 per cent. 'A deep recession and large reconstruction costs are to be expected, on the backdrop of a humanitarian crisis' (BBC, 2022). Clearly, rebuilding Ukraine will be financially more expensive than conducting the war itself. The World Bank too estimated, in 2022, that the cost of rebuilding Ukraine, when the war ended, would be about $349 billion, which is larger than Ukraine's pre-invasion GDP and three times greater than all the military, humanitarian and financial assistance commitments to Ukraine put together since the start of the war. Two years have passed since this estimate, and this figure will certainly be much higher when the war does finally come to an end. The Kyiv School of Economics provided the chart in Table 3.1 for the total damage in monetary terms, measured by 1 September 2023.

The impact and consequences of the war are far more complex than just military and economic costs. The main losers of the war and all this damage to infrastructure are Ukraine's civilian population, those living both in the Russian-controlled eastern provinces and in the rest of the country controlled by the Ukrainian government. More than anything else the war is seriously damaging human security. The war poses an existential risk not only to the Ukrainian state but also to its people's living conditions. The war so far has caused the deaths of thousands of civilians, displaced millions and severely damaged the economy, houses, schools, hospitals, cultural sites and natural environment in the country.

Unfortunately, the people of Ukraine will feel the serious consequences of the war long after the guns eventually fall silent. Probably several decades will pass before the reconstruction efforts are finalised. The war is, more than anything else, a human security catastrophe. The 2022 war initiated by Russia has unleashed the greatest human security crisis in Europe since World War II by seriously threatening the safety, livelihoods and dignity of more than 36 million people living in Ukraine and their communities. In the words of Paul Rogers, 'the war is also having an

Table 3.1 Total estimate of infrastructure damage by industry in monetary terms, as of 1 September 2023

Property type	Damage, $ billion
Housing	55.9
Infrastructure	36.6
Assets of enterprises, industry	11.4
Education	10.1
Energy	8.8
Agriculture and land resources	8.7
Forests	4.5
Transport	3.1
Healthcare	2.9
Utilities	2.7
Trade	2.6
Culture, sport, tourism	2.4
Administrative buildings	0.5
Digital infrastructure	0.5
Social sphere	0.2
Financial sector	0.04
Total	**151.2**

Source: KSE, 2023.

impact on global human security, directly in the form of food shortages and indirectly in its economic consequences. These are likely to get significantly worse from the autumn if the war grinds on' (Rogers, 2022). In liberal democratic societies, human security and state security are interconnected components that complement each other. Only those societies can be resilient, where people are properly protected from the entire range of threats, and where human security is guaranteed in its modern and inclusive sense (Khylko and Tytarchuk, 2017: 1).

Putin started this war expecting a quick and easy victory. He underestimated the resolve of Ukrainians to fight. Ukraine (and NATO) also overestimated its capacity to defeat Russia on the battlefield. Ukraine has already suffered levels of damage not seen in Europe since World War II. The war has so far destroyed residential buildings, educational institutions and the infrastructure and industry sector, leading to a huge level of overall damage. Daniel Treisman identifies four schools of thought regarding Putin's decision to invade Crimea, all of which could apply to his decision to invade Ukraine: 'Putin the defender', responding to the potential for Ukraine to join NATO; 'Putin the imperialist', seizing Crimea as part of a broader project to recreate the Soviet Union; 'Putin the populist', using the annexation/invasion to build public support; and 'Putin the improviser', seizing a fantastic opportunity with well-prepared military operations, but chaotic political arrangements (Treisman, 2018). We can add a fifth: 'Putin the war criminal'.

Chapter 4 continues and expands the discussion on war crimes, tribunals, accountability and the future of warfare.

Notes

1 The decision of the Russian government to recognize the independence of the two separatist republics on 21 February 2022 effectively put an end to discussions of Minsk implementation. Three days later, Russia invaded Ukraine, heralding a new phase of the war.
2 Eurasianism emerged in the 1920s among Russian émigrés and, though pretty much unknown to the public inside the Soviet Union at the time, became a quite fashionable creed in the late Soviet and post-Soviet era. Eurasianists, despite all of their differences, had a common ideological denominator: all of them believed that Russia belongs neither to the East nor to the West, but it is a civilisation in its own right. Eurasianists also believed that Russia emerged as a peculiar blend of Slavic and mostly Turkic people. They believed that Russia/Eurasia is closer to the East, from a civilisational point of view, than to the West, even though some post-WW II Eurasianists incorporate part of Europe (notably Germany and France) in Eurasian cultural/geopolitical space. (Shlapentokh, 2009).
3 The choice of the venue was not coincidental. The skirmishes between the Russian forces and German knights took place at this location in the 13th century, which leads to Russians consider Izborsk as Russia's 'western outpost'.

4   War crimes and future wars

This chapter revisits war crimes, in light of the recent trials in Ukraine, starting with the conviction of 21-year-old Russian soldier Vadim Shishimarin, on 23 May 2022, who was sentenced to life in prison for killing an unarmed Ukrainian civilian by a court in Kyiv. A week later, a Ukrainian court sentenced two captured Russian soldiers to more than 11 years in jail, each for shelling a civilian area in the country's east amid Moscow's offensives. The sentencing of Alexander Bobikin and Alexander Ivanov was the second verdict handed down in war crimes trials held by Ukraine since the start of Russia's invasion in late February 2022.

What war crimes have been committed in Ukraine and what determines which perpetrators of violence are held accountable and which are not? What does this mean for international law, justice and accountability, as well as for the continuation of acts of aggression against civilians by states and by non-state actors? Moreover, what role does definitional bias, when it comes to what constitutes a 'legitimate target', a 'war crime' and a 'civilian', play in hindering impartiality and justice, whether through the International Criminal Court (ICC) or through partisan courts? We further ask if there is a type of hard warfare that avoids the commission of war crimes, civilian harm and humanitarian catastrophe. To this end, the chapter evaluates remote warfare and its impact, paying particular attention to war and ethics in light of the new technologies used in both wars under study, this time in the context of Ukraine. Debates about the rules and ethics of war have taken place for thousands of years, but those debates have relevance and new applications for theorists and practitioners today, as technology enables more and more ways to fight enemies: drones, satellites, computers and global positioning systems (GPS). As remote warfare is becoming more popular, and employed by governments and militaries in civil, defensive and aggressive wars, is war becoming cleaner, moral and just? As the world moves towards a multi-polar security system, are imperial wars increasing perceptions of persisting bipolarity, or (depending on the outcome of the Ukraine war) even unipolarity? How would that then affect global security dynamics and further human costs?

The chapter concludes with an assessment of the present and future of international justice. The war in Ukraine has shown that justice and accountability,

DOI: 10.4324/9781003414827-5

when it comes to crimes against civilians, are possible, even in an ongoing conflict. But which/whose crimes remain unpunished?

War crimes revisited

Sergeant Vadim Shishimarin shot a 62-year-old Ukrainian man through an open car window in the eastern village of Chupakhivka. When he and four other Russian soldiers arrived in Chupakhivka, about 200 miles east of Kyiv, they saw a man cycling and talking on his phone. Shishimarin was ordered by another soldier to kill the civilian in order to prevent him from raising the alarm that Russians were in the village. Shishimarin then killed the man just metres from his home, 'firing an assault rifle several times through the open window of the car at his head' (Russell, 2022).

By the first anniversary of the invasion of Ukraine by Russia, 25 Russian soldiers had been convicted of war crimes in Ukrainian courts, including 'a soldier who forced two Ukrainians at gunpoint to hand over laptops and money, four who beat and tortured Ukrainian soldiers, and two who admitted shelling residential buildings in the first weeks of the war' (Sly, 2023). Over 66,000 additional alleged war crimes have been reported to Ukrainian authorities, according to Ukraine's Office of the Prosecutor General. The growing number of complaints ranges from the theft of property to torture, murder, rape, the deportation of Ukrainian children to Russia and missile strikes against Ukrainian infrastructure. Ukraine's prosecutor general, Andriy Kostin, has vowed to investigate all of them and to bring those responsible to trial. As for President Zelensky, he has made justice for the victims of war crimes one of his conditions for eventual peace with Russia.

The impact of this war on Ukrainians – the warriors and the non-combatants – has been well documented. From 24 February 2022, which marked the start of the large-scale armed attack by Russia, to 8 October 2023, the UN Human Rights Office recorded 9,806 civilians killed (UN Office of the High Commissioner for Human Rights, 2023). Explosive weapons accounted for 90 per cent of all civilian casualties.

Reminiscent of the Iraqi dead, the names of those lost were entered onto an online memorial, trying to contain the immense loss of a mother, a grandmother, a father, a son, inside a few words.

Oleksii Bondar was killed on March 19, 2022, in Mykolaiv. The Russian army carried out an airstrike on his family's house. The 7-year-old boy died with his mother and grandmother. Oleksii had just started school. He liked to play football.

(Memorial platform tweet, 16 May 2023)

85-year-old Svitlana Larionova died with her daughter Iryna on the night of March 2, 2023. A Russian missile hit their home in Zaporizhzhia. Svitlana's body was recognised by the earrings that her daughter gave her for her birthday.

(Memorial platform tweet, 14 June 2023)

22-year-old Mykola Klontsak was killed on March 11, 2023, in the battle for Bakhmut. 'He had been working since he was 14. He grew up without a father, his father died when Mykola was 11 years old. I always knew I could count on him. Now he is our Angel,' – says his mother.

(Memorial platform tweet, 3 June 2023)

Many of those now fighting – and dying – in defence of their country were civilians when the war had started:

22-year-old Vasyl Kosovskyi was killed in action on March 16, 2023, in Donetsk. Vasyl lived in Sumy Region. He worked as a milk truck driver and a car mechanic.

(Memorial platform tweet, 26 May 2023)

Much like the wars in the Middle East, this war has already taken tens of thousands of lives, with over 100,000 estimated killings of civilians and combatants, Ukrainians and Russians, in fewer than two years.

In his book, *Invasion: Russia's Bloody War and Ukraine's Fight for Survival,* Harding adopts a self-defence approach when narrating and analysing the war. The clear aggressor and instigator of the war is an imperialist Russia. He writes:

Putin had issued a series of demands so imperious and swaggering you could only marvel at their audacity. He sought nothing less than the annulment of the security infrastructure that has governed Europe for the three decades since the Soviet Union's 1991 collapse. Further, he wanted the Biden administration to guarantee Ukraine would never join NATO, the United States-led military alliance set up in 1949 to contain the Soviet Union.

(Harding, 2022: 4–5)

Before invading Ukraine on 24 February 2022, Putin also demanded that NATO take its forces and equipment out of Romania, Bulgaria, Poland, Latvia, Estonia and Lithuania – former Eastern Bloc countries that had joined NATO after 1997. Putin's goal, Harding argues, was to recreate the Soviet Union's sphere of influence that had existed across the European continent behind the 'iron curtain'. This area 'encompassed Belarus and Ukraine – "historic" Russian lands, as Putin saw them – unjustly separated from Moscow by Bolshevik blunder and Western meddling' (Harding, 2022: 5).

Putin used tactics familiar from Russia's dark past, both distant and recent: bombs, destruction and the killing of civilians. This time Russia's direct enemy was Ukraine, its indirect enemy was its Atlanticist leadership. In Russia's attempts to create a new world order and become a global hegemon, this invasion would become the largest conflict in Europe since 1945: 'an attempt by one nation to devour another' (Harding, 2022: 6). One expert even described it as Russia's strategy to wipe out 'Ukraine-ness' (Harding, 2022: 106), while simultaneously fighting a proxy war against the West. And it was a West-backed Ukraine that responded to the Russian aggression. Within days of the invasion, Sweden and Finland abandoned neutrality, Germany announced a radical shift in its security policy, and the USA and its allies found a new role and moral purpose in their solidarity – both moral and material – with Ukraine.

It was also a defensive war agenda that the words of Ukraine's president betrayed from the start. Zelensky's presidential term began on 20 May 2019, almost three years before the invasion. Referring to the Russian annexation of Crimea and the Russian-instigated conflict in Donbas, in his inaugural address to the Ukrainian Parliament, he declared: 'I am ready to pay any price to stop the deaths of our heroes' and secure 'the "return" of the lost territories…Crimea and Donbas are Ukrainian land' (Zelensky, 2022: 15–16). He soon embarked on a quest to find willing and powerful allies to help him defend Ukraine.

- Zelensky's address to the UN General Assembly, New York, 25 September 2019: 'What is happening in my country is no longer "someone else's war". None of you can feel safe when there is a war in Ukraine. Not when there is a war in Europe' (Zelensky, 2022: 19). 'Military methods, technologies and weapons mean that our planet is not as large as it once was', he pointed out, adding, 'A strong leader is the one who protects the lives of everyone' (Zelensky, 2022: 22).
- Zelensky's address to the US Holocaust Memorial Museum, Washington DC, 1 September 2021: 'Do not be indifferent to Ukraine' (Zelensky, 2022: 31).
- Zelensky's speech to the Munich Security Conference, 19 February 2022: 'We appreciate any help, whether it is hundreds of modern weapons or 5,000 helmets' (Zelensky, 2022: 45).

The day after Russia invaded, Zelensky addressed the people of Europe from Kyiv:

I know that Europe can see this. But what we do not see – at least not fully – is what you are going to do about it … to protect Ukraine… When bombs fall in Kyiv, they fall in Europe. When missiles kill Ukrainians, they kill Europeans. More protection for Ukraine means more protection for the democratic world.

(Zelensky, 2022: 58–59)

Appearing remotely in the German Bundestag, three weeks after the invasion, he urged Germany to 'Support us. Support peace'. He contended that Ukraine had always striven for peace and had always been committed to dialogue, to negotiations and to a diplomatic solution. Yet on 24 February 2022, at 4:30 a.m., Ukraine had received Russia's answer, made clear through its actions: Russia wanted 'to destroy Ukraine; to wipe us off the face of the Earth, both as a state and as a people', said Zelensky (Zelensky, 2022: 5).

Zelensky's efforts are another example of leader responsibility as an important factor at the start of every war. Like other leaders before and since, in this and in other wars, he could have made different decisions that would not have resulted in a war that has taken the lives of so many people, civilians and combatants.

Recording the casualties

There are several official and unofficial organisations recording the devastating impact of Russia's war on the people of Ukraine. Action on Armed Violence (AOAV) has created statistics using English media reports. There are currently two Ukrainian organisations collecting information about casualties: Children of War, created by the Ukrainian government to show pictures and names of children who have been killed, displaced/missing or injured by Russian aggression; and Memorial Platform, which posts names, faces, ages and circumstances in which the person died. The data are collected through a website where someone can send information about who has been killed, Ukrainian and other nationals, civilian or military.

AOAV, launched in October 2010, collects news reports of explosive violence from English-language media to create and update reports about civilians killed or injured by explosive weapons around the world. Between February 2022 and November 2023, AOAV recorded 5,202 civilian deaths in Ukraine (Action on Armed Violence, 2023). AOAV states that this figure should be considered the lowest estimate, as the information is very difficult to collect and verify. Among the civilian casualties, AOAV has recorded 627 Ukrainian children. Ground-launched explosive weapons (missile strikes, artillery shelling and rockets) have caused 71 per cent (12, 362) of civilian casualties, while the vast majority of the locations where civilian casualties were recorded were urban residential areas, including town centres, schools, hospitals, hotels and commercial premises.

The Human Rights Monitoring Mission in Ukraine (HRMMU) has been recording data from Ukraine since 2014, when Russia annexed Crimea. Since 24 February 2022, the HRMMU has been focused on 'documenting violations of international human rights law and international humanitarian law committed by all parties' (OHCHR, 2022). The data collected are on the deaths and injuries of civilians, while also documenting the sex, age, place of incident, type of incident and weapon used. Public updates are released frequently to show that the information is continuously being collected, which creates transparency and consistency (Ward, 2023). The reports on civilian casualties are collected through fieldwork, which includes interviews with victims or families and witnesses to events. The information collected is verified with the aid of official records, open-source documents, photo and video materials, forensic records, criminal investigation materials, court documents, reports by international and national non-governmental organisations, public reports by law enforcement and military actors and data from medical facilities and local authorities. There is a disclaimer that current statistics that the HRMMU is currently producing are based on individual civilian casualty records where there are 'reasonable grounds to believe' as a standard of proof (UNHR, 2023). The main reason the UN collects data on casualties is to show the disregard for human rights (Ward, 2023). Matilda Bogner, Head of the Mission in April 2022, stated that fact-finding missions can provide justice for victims and hold perpetrators accountable. Missions like this can also apply pressure to prevent further violations from being committed (UNHR, 2022). Bogner stressed that the people collecting the data need to have a victim-centred approach, saying 'we need to respect the victims. We need to respect the witnesses' wishes. If they don't wish to talk (...) we respect their wishes and don't further stigmatize them' (2022). But while the UN is transparent about why it collects the data and regularly updates its figures, inclusivity and consistency are hard to see in the UNHR and HRMMU reports. Ward argues that, without giving details, such as names or causes of death (other than from explosive violence), it is hard to see what is left out or has been removed from the statistics.

There are many Ukrainians who want to share their stories of family and friends lost in this conflict, and a couple of organisations have decided to document and humanise the deaths by showing the faces and names of those Ukrainian adults and children who have died as result of the Russian aggression. Memorial Platform is one of those organisations, releasing images, names, ages and locations, documenting and showing the faces of those who have been killed: military personnel, civilians and humanitarian workers from any nation. The platform provides two links for people to submit information on either civilian – 'civilian victims' – or military deaths –'fallen heroes' – by filling out an online form. Data are also collected by journalists of partner organisations working in the field. Similar to AOAV, they monitor social media and media mentions of any victims or large-scale events. There are three categories of deceased: military, civilian and children. There are three additional pages dedicated to writings created by victims/ journalists to express additional memorial texts about victims/sites or impactful events. Memorial was launched in March 2022, and one of the victims whose information was posted was Liza Dmitrieva, a 4-year-old girl with Down's syndrome, who died just outside her developmental speech therapy class in Vinnytsia as a result of a Russian air strike. She was being pushed in a stroller through a crowded square and was killed alongside 23 other people. The aim of the page is to 'help preserve the memory of everyone whose life was cut short by Russia's war against Ukraine' (Memorial Platform, 2022). Photos are accompanied by a brief description of the deceased to show appreciation for their life before they were suddenly and unjustly killed.

Fallen heroes commemorated on the Memorial Platform include Dmytro Tsis.

On September 26, 2023, Dmytro Tsis, a sapper of the State Emergency Service, died in the Burn Center of Kyiv City Hospital No. 2. On September 15, during demining of a forest massif near the city of Lyman in Donetsk region, the car in which Dmytro and his colleagues were traveling ran into an anti-tank mine. The man received severe burns to his face and upper limbs. Doctors fought for his life for 11 days, but it was not possible to save him. Dmytro was 24 years old. He lived in Poltava. He worked as a sapper of the department of pyrotechnic works of the State Emergency Service of Poltava region, and was a sergeant of the civil defense service. He studied at the Poltava Educational Complex No. 16, was a classmate of my best friend. Many girls liked him. Had an incredible smile ... Tried to be useful, was the soul of the company. He liked to play basketball. When we were teenagers, we had a big enough company, gathered after school, played football, Dracula, basketball. Dima tried to play everywhere ... He was very intelligent, shrewd, principled ... He would still like to live and live ..." said Tatyana, Dmytro's acquaintance.

(Memorial Platform, 2023)

He is survived by his father.

Memorial Platform is a non-governmental, non-commercial initiative owned by the Abo media growth agency, which has a network of media partners across Ukraine. The two goals of the Memorial Platform have been to humanise statistical figures and to form a strong institution of national memory as one of the prerequisites of security in the future. Memorial Platform has a full-time team of over ten people and dozens of freelance authors working in different regions of Ukraine. All are experienced journalists or have worked as war correspondents. By the first anniversary of the invasion, they had collected information on 2,300 victims of the conflict, civilian and military. Memorial Platform, like Iraq Body Count and AOAV, has been a member of the Casualty Recorders Network since February 2023.

Children of War was created by the Ministry of Reintegration and the National Information Bureau on behalf of the Office of the President of Ukraine not only to document the deaths of children but also to help find and rescue children who have been displaced or deported to Russia. It is the only platform that provides up-to-date information surrounding children who have suffered or are suffering because of the Russian invasion. With the assistance of law enforcement agencies, the national police of Ukraine, the office of the prosecutor general and the national information bureau, they have been able to determine the following for the period 24 February 2022–5 November 2023: 510 children dead and 1,143 wounded; 2,024 missing and 20,088 found; 19,546 deported/displaced; 13 sexually abused (Children of War, 2023). On this platform, people can report a lost child, a crime committed against a child, a child found unaccompanied, a deported child or any violation of a child's rights and freedoms during the war.

Children of War also contains children's stories, 'the stories of children affected by the full-scale war', based on interviews conducted under the guidance of child psychologists and by professional teams 'with maximum consideration of the child's interests'. Each child's story is a testament to the trauma suffered by survivors. These are two such stories.

Bohdan, 8 years old, Bakhmut. His parents were killed by an enemy shell right on the street. The boy's mother was seven months pregnant. Bohdan was staying with a neighbor at the time, and when everyone fell asleep, he took his bike at night and went to the scene of that terrible event to see his parents. His little heart refused to believe that his mom and dad were gone. The bodies of his parents remained lying in the street for several days. Because of the constant shelling, it was not possible to bury them immediately. Law enforcement officers rescued Bohdan. They came to the boy, gathered his belongings, evacuated him from the hot spot, fed and entertained him. The boy was scared and desperate after hearing the news of his parents' death.

Dima and Anya, 12 years old, Chernihiv Region. Dima and Anya from the Yahidne village, Chernihivska oblast, are friends and neighbors. They were among the 380 people who were rounded up by the occupiers and kept in the basement. 'Yahidne is a small village, there are five streets and everyone knows each other,' says Anya. 'There were only three kids in my class,' adds Dima. The Russians set up headquarters in the local school, and all the villagers were rounded up into the basement. 380 people spent 28 days in the cramped basement without the ability to move freely. 'They were having fun when they shot at us and we were scared.' Instead of a restroom, there was one bucket for everyone. A list of people who died of suffocation or were shot was written on the wall. Dima's mother went up to the Russians and asked them to give her a piece of bread or shoot her. It was unbearable to watch her children starve.

(Children of War, Children's Stories, 2023)

A 'useful information' section alerts the visitor to the dangers, risks and suffering that war brings to everyone, including the children. The page contains a variety of documents on what to do if a child is deported, if you know about war crimes committed by Russians against children and if you want to report a war crime. The portal also provides security rules during wartime and offers advice on what to do if you detect a suspicious object, instructions on self-protection in emergency situations and a booklet for people who have met/discovered an unaccompanied child. There are instructions for doctors, a video course on how to shelter a child in the family, a guide on how to contact institutions that provide assistance under martial law, how to register children who have been evacuated and how to support a child when dad/mom is at war. Moreover, there is guidance for foreigners willing to shelter children from Ukraine.

These organisations give us glimpses of the death, the pain and the loss suffered by millions in wartime. They provide a remembrance of lives lost, of lives endangered, of lives disrupted, and also of crimes committed by a large aggressive state.

War and ethics: Ukraine's just war

The just war tradition is the predominant moral language through which we address the rights and wrongs of the use of force in international society. It provides us with a set of concepts and principles for understanding and responding to the moral and legal questions raised by war. As such, it is central to the theory and the practice of international relations. Its influence is evident in the legal codes that govern how modern militaries perform their duties, and it has featured prominently in the rhetoric surrounding the War on Terror and the military actions in Iraq, Afghanistan and Libya (Lang, O'Driscoll and Williams, 2013: 1). Just war adopts a universalist perspective in its analysis and moral understanding of security and war within the context of an international community that both embraces and transcends all states, recognising *mutual* rights and obligations where realists argue that states only have duties to themselves.

The universalist/international perspective of just war theory can be seen in current international law, especially the laws of armed conflict. Just war theory and the laws of armed conflict apply abstract universal rules of conduct to warfare and hold the same core value convictions that war is sometimes permissible, and there's a difference between permissible and impermissible means of fighting it. While just war theory is much older, drawing its values and understandings from religious writings, ethical values, political debates and military experiences, the laws of armed conflict were constructed in the modern era by national governments, turning to just war theory for guidance (Orend, 2019: 80–81). 'Traditional just war theory maintains that the right to go to war is an integral part of international law' (Williams, 2012: 44).

The laws of armed conflict stipulate that a state may resort to armed force if there is just cause, if there is proportionality in its response and if the use of force is a last resort.

Just Cause is the most important rule that sets the tone for everything else that follows. When it comes to a just cause for going to war, international law (UN Charter articles 39–53) recognises three general principles:

- All countries have the inherent, 'natural' right to go to war in self-defence.
- All countries have the inherent, 'natural' right of other-defence – to go to war as an act of aid to any country victimised by aggression.
- Any other use of force is not in the eyes of international law an inherent, 'natural' right of states. Any country wishing to engage in more controversial forms of force (such as a pre-emptive strike) needs the approval of the UN Security Council.

Proportionality: The laws of armed conflict command that the problem in question be serious enough that war is a proper, fitting reply. So what problem is so severe that a declaration of war is justified? International law's answer is *aggression*:

When confronted with an aggressive invader (…) it's deemed reasonable to stand up to such a dark threat to life and liberty and to resist it and beat it back, with force if need be. Just as dangerous criminals must be resisted and prevented from getting away with their crimes – lest chaos be invited – countries are entitled to stand up to aggressors, and to defeat them.

(Orend, 2019: 86–87)

How are belligerents supposed to fight? It can be argued that once war has begun, in theory, everything is permitted; in practice, it may be wise to restrain oneself. It depends on the nature of the enemy, the level of threat, their intention and capability, one's own military capability and what's more likely to contribute to one's objective in the circumstances. Both just war theory and the laws of armed conflict impose firm permissions and prohibitions on the conduct of belligerents. The overall function of *jus in bello* is (1) to uphold a kind of ideal regarding a fair and decent way to fight, and (2) to put clean limits on the amount and kind of force being deployed, in particular, so that so-called 'total war' – i.e. unlimited, indiscriminate escalation in violence – does not happen (Orend, 2019: 112). Without limits or restrictions, massacres, ethnic cleansing, genocide or the use of WMD may occur. Hundreds of laws of armed conflict serve to realise six principles of just war fighting:

1 Discrimination and non-combatant immunity.
2 Benevolent quarantine for prisoners of war.
3. Proportionality.
4. No use of prohibited weapons.
5. No use of means 'mala in se' ('evil in itself' – acts inherently immoral, like murder, or rape).
6. No reprisals.

The intentional killing of civilians is considered the worst war crime.

Ukraine, as we have so far argued, has been fighting a war of self-defence against Russia. The ethics of self-defence have implications for the moral justification of national defence, whereby self-defensive behaviour, whether by individuals or by nations, is morally superior to aggressive behaviour. 'Self-defence thus possesses an aura of respectability that seems to render it eminently suitable as a rallying point for agreement on the ethical legitimacy of warfare' (De Roose, 1990: 159). Killing a person in response to a threat to one's life, as in defensive warfare, is regarded as a morally justified self-defensive act: 'killing in self-defence when that is the only option available for saving one's life seems to be the pre-eminent instance of morally justified lethal self-defence' (De Roose, 1990: 159). On the other hand, killing when it is done by aggressors is not morally justified; the aggressor's behaviour is morally wrong, while the victim's (aggressive) response is morally right, because of the victim's moral right to protect himself/herself from deadly aggression. There is then a presumption of the aggressor's culpability and the victim's innocence/ lack of culpability. By killing an aggressor, victims of violence, especially when lethal force is used, do not incur any responsibility for having killed someone but merely make the aggressor take responsibility for his or her act, by letting him or her bear the act's inevitable effects. 'As long as victims are in a situation in which they cannot but choose between killing their aggressors and being killed by them, they are under no moral obligation to refrain from killing their aggressors' (De Roose, 1990: 162).

The principle of morally justified self-defensive killing in war implies that aggressive enemy forces in uniform can be killed by other combatants/soldiers fighting a defensive war, unless they are surrendering or are *hors de combat*, because the lives of the soldiers defending their country and the lives of their co-nationals are threatened by the enemy forces. In this case, which is the case made for Ukraine, the imminent threat to the lives of those on the battlefield can only be averted by killing the enemy. Morally, defensive military action is very different from military aggression, with the main goals of defensive military action being to ward off aggression, to save lives and to limit the bloodshed, all through the use of force.

Any human cost (especially civilian) resulting from Russia's current war of aggression is both immoral and illegal. It is a war crime. But what if this is not a just war? What has been presented as a Western effort to help Ukraine defend itself from Russian aggression can also be interpreted as a campaign to weaken Russia and degrade its military capacity to fight wars. In March 2022, one month after the start of the war, Chas Freeman, a retired American diplomat, commented in an interview:

> Everything we are doing, rather than accelerating an end to the fighting and some compromise, seems to be aimed at prolonging the fighting, assisting the Ukrainian resistance -which is a noble cause, I suppose, but … will result in a lot of dead Ukrainians as well as dead Russians.
>
> (Freeman, 2022)

To weaken Russia, a long and bloody war would be required, with many Russian and Ukrainian casualties, but no American. In Freeman's words, America will feed and prolong this war 'to the last Ukrainian for Ukrainian independence', sparing neither combatants nor civilians. Ukraine will too. All this is exactly what Zbigniew Brzezinski, former US national security adviser, famously argued in his 1997 book, *The Grand Chessboard*, that neither the West nor Russia can afford to lose Ukraine to its strategic and economic adversary. 'If Moscow regains control over Ukraine, with its 52 million people and major resources as well as access to the Black Sea, Russia automatically again regains the wherewithal to become a powerful imperial state, spanning Europe and Asia' (Brzezinski, 1997:46). The loss of Ukraine was geopolitically pivotal, for it drastically limited Russia's geostrategic options. According to the Western narrative, Putin is an expansionist, an evil man, a new Hitler, not motivated by national security concerns but by aggression similar to that of the Nazis in World War II. However, there is a different context in which Russia's violence can be understood. Since the end of the Cold War, the USA and it European allies have expanded NATO over a thousand miles eastward, despite assurances that they would not do so; withdrawn from the Anti-Ballistic Missile Treaty and placed anti-ballistic launch systems that can fire nuclear weapons (such as nuclear-tipped Tomahawk cruise missiles) at Russia in former East European, and now new NATO, states; conducted NATO military exercises near the Russian border; started NATO membership negotiations with Ukraine;

armed and trained Ukraine's forces. The US Congressional Research Service reveals that between 2014 and early 2022, US security assistance to Ukraine amounted to over $4 billion, most of which came from the Department of Defense and the State Department. This amount has increased to $42 billion following the Russian invasion. This assistance has included 'logistics support, supplies, and services; salaries and stipends; sustainment; weapons replacement; and intelligence support'. Through the Joint Multinational Training Group-Ukraine, which was established in 2015, the US Army and military trainers from US allies, 'provided training, mentoring, and doctrinal assistance to the UAF before the war'. In 2017, the Trump Administration announced that the USA would provide lethal weapons to Ukraine: 'rocket-propelled grenade launchers, counter-artillery radars, Mark VI patrol boats, electronic warfare detection and secure communications, satellite imagery and analysis capability, counter-unmanned aerial systems (UAS)' (Congressional Research Service, 2023).

The anti-ballistic-missile (ABM) system that the USA put into operation in Romania in 2016 uses the Mark-41 'Aegis' missile launchers, which can accommodate Tomahawks with a range of 1,500 miles, capable of striking Moscow and other targets inside Russia. Tomahawks 'can carry hydrogen bomb warheads with selectable yields up to 150 kilotons, roughly ten times that of the atomic bomb that destroyed Hiroshima' (Abelow, 2022: 20). A similar site has been constructed in Poland. John Mearsheimer noted:

> Other NATO countries got in on the act, shipping weapons to Ukraine, training its armed forces and allowing it to participate in joint air and naval exercises. In July 2021, Ukraine and America co-hosted a major naval exercise in the Black Sea region involving navies from 32 countries. Operation Sea Breeze almost provoked Russia to fire at a British naval destroyer that deliberately entered what Russia considers its territorial waters.
>
> (Mearsheimer, 2022)

In 2020, NATO conducted a live-fire training exercise in Estonia, 70 miles from Russia's border, using missiles with ranges up to 185 miles. The following year, NATO fired 24 rockets, simulating an attack on air defence targets. In June 2021, in the meeting of the North Atlantic Council in Brussels, NATO reaffirmed its commitment: 'We reiterate the decision made at the 2008 Bucharest Summit that Ukraine will become a member of the Alliance' (Brussels Summit Communiqué, 2021: para 69). Less than two months before the invasion of Ukraine, on 30 December 2021, Anatoly Antonov, Russian ambassador to the USA, warned that 'everything has its limits', and that if NATO kept 'constructing military-strategic realities imperilling the existence of our country', Russia would be forced 'to create similar vulnerabilities for them. We have come to the point when we have no room to retreat. Military exploration of Ukraine by NATO member states is an existential threat for Russia' (Antonov, 2021). Russia could then also argue that it is fighting a defensive war due to the existential threat posed by (1) a Western-armed, trained and militarily integrated Ukraine, and (2) by NATO.

Who is responsible for the humanitarian disaster in Ukraine, for the killing and wounding of hundreds of thousands of civilians and soldiers, Ukrainian and Russian? Who bears responsibility for the destruction of homes and the creation of yet another refugee crisis? Who has contributed to the ongoing harm that is being inflicted on the economies of European countries? And who will bear responsibility if the war becomes nuclear?

The obvious answer is Russia, because President Putin declared this war and is directing its conduct. However, Russia is not the only responsible party, as President Zelensky's acts can be seen as threatening and causing the war by provoking the invasion. By actively seeking Western support (financial, material and military) since his election in 2019, he can be said to have been acting, to use Suganami's term, 'thoughtlessly', knowing how his actions were perceived by Russia, but not giving serious consideration to the resulting risk of war. It could be argued that he has acted 'recklessly', calculating that committing the offensive/threatening acts would lead to more gains, although the probability of war resulting from his acts and its cost would be considerable. Acts carried out after such calculation, risking war in pursuit of long-term gains, are 'reckless' acts (Suganami, 1996). This leads to the conclusion that Ukraine's leader acted thoughtlessly and/or recklessly and bears some of the responsibility for the war being fought on Ukrainian soil and for its casualties.

The week before the Russian invasion, Zelensky met with German Chancellor Olaf Scholz in Munich. During this meeting, the German Chancellor offered to broker a peace deal between Ukraine and Russia, telling Zelensky that if Ukraine declared neutrality by not joining NATO, the security of Ukrainians could be achieved.

> Ukraine should renounce its NATO aspirations and declare neutrality as part of a winder European security deal between the West and Russia. The pact would be signed by Mr Putin and Mr Biden, who would jointly guarantee Ukraine's security. Mr Zelensky said Mr Putin couldn't be trusted to uphold such an agreement and that most Ukrainians wanted to join NATO.
>
> (Gordon, Pancevski, Bisserbe and Walker, 2022)

Had this been a war between two states, a weaker one and a stronger one – Russia and Ukraine – similar to a war, say, between the USA and Iraq or Afghanistan, then the probability of the weaker state winning the war would be almost zero. Iraq and Afghanistan fell within a few weeks, having had no support from any other state and having been attacked by a hegemonic power. Ukraine would have probably suffered the same fate, if it had had to defend itself, unsupported, against Russia. What gave the Ukrainian leader confidence in a high probability of victory in a war against Russia was the backing of NATO states. However, what that has enabled is a long war that raises the human cost daily.

If this is not a war between two states, the responsibility of NATO needs to be taken into account too; NATO/US assistance has been focused on providing capabilities that Ukraine's domestic defence industry cannot produce, increasing its ability to sustain offensive and defensive operations, thus prolonging the war. Provisions have included drones, advanced rocket and missile systems, communication and intelligence support, battle tanks, fighter aircraft and additional air defence capabilities. The provision of security assistance is also focused on improving its 'medium- to long-term capabilities, including transitioning towards more NATO-standard weaponry' (Congressional Research Service, 2023). With a large anti-Russian organisation behind it and an aggressive hegemonic state arming it, the war Ukraine is fighting looks a lot less like a war of self-defence, with Russia as the only aggressor and the Russian leader bearing full responsibility for the horrors the world has witnessed since February 2022.

Instead of supporting a negotiated peace in the Donbas region between Kyiv and pro-Russian autonomists, the US government encouraged strong nationalistic forces and poured weapons into Ukraine, which led Ukraine to adopt intransigent positions towards Russia. It can then be argued that the war being fought in Ukraine is a US/NATO remote war involving distance and delegation, in other words, a proxy war, or a war of imperialism, which, according to just war theory, is an unjust war. 'Russia's battle went beyond Ukraine. It was – to a large degree – proxy war against the West', writes Harding (2022: 7). Yet the West, through Ukraine, has also been fighting a proxy war against Russia. In this conflict, both Russia and the West found a new moral purpose that, yet again, has involved militaries and weapons. Does that mean that Russia is 'off the moral hook'? In just war theory, Russia is still the belligerent, the aggressor, so Russia's killings are still immoral and criminal. But the moral responsibility for the human costs of this war – the dead, the homeless, the displaced and the abused – lies with more than just the Russians.

Remote warfare in Ukraine

The War on Terror has shown how remote warfare both stretches and shortens the battlefield, as the USA launched risk-free drone strikes in Afghanistan, Iraq, Pakistan and Yemen. Those 'precision'-targeted assassinations sometimes hit the targets – those individuals they hunt – and sometimes they don't, with devastating consequences. In Ukraine, both sides use drones for a range of missions, including 'battlefield surveillance, artillery spotting and attacking armoured vehicles and missile launchers' (Gonzalez, 2023). The Ukraine–Russia conflict is carried out on a hypermodern battleground where, more than ever before, drones foreshadow a world in which armed conflicts could end up being conducted exclusively by remote control and even fully by artificial intelligence. The war in Ukraine resembles past conflicts that involve the use of fighter planes, tanks, helicopters, anti-aircraft missiles and warships, but in addition to more traditional warfare, the Russian invasion of Ukraine has led to a high-intensity war where both sides have extensively deployed military and commercial drones as an extension of air power or as ammunition (Kunertova, 2023).

What the war in Ukraine has highlighted is the breadth and width of the international arms trade, when it comes to the export and supply of lethal weapons in states worldwide. In particular, following the end of the Cold War, weapons dealers from Western states targeted former Eastern bloc countries, as they looked to make a profit on weapons systems (Stohl and Grillot, 2009). What we also see in the arms trade though is its morally and legally indiscriminate character. Let us take the UK as an example. The UK is a leading member of NATO, and the UN is a democratic state and human rights proponent. Yet the UK company Mil-Tec Corporation Ltd supplied $6.5 million worth of weapons to the Hutu regime in Rwanda to forces committing genocide, at least from June 1993 to mid-July 1994. Moreover, in the 21st century, the military development and production complex is 'sufficiently integrated and powerful to have considerable influence in determining international security policy. It embodies an outlook that prioritises military responses to perceived threats' (Rogers, 2021: 145). The modern-day military-industrial complex exists globally, but especially in powerful states such as the US, Russia, China, the UK and France. Each military-industrial complex has interconnected components: arms companies, senior military personnel, civil servants, arms sales, profitable enterprises and a hunt for new enemies, 'all adding to a culture that can always depend on appeals to patriotism' (Rogers, 2021: 145). The whole complex exists as part of a wider security culture, one that privileges national interests and favours military responses to presumed threats. It is a culture of violence, a global culture of war, within which victims are collateral damage and moral values are of little relevance. When it comes to the world's arms dealers, 'their primary function, like that of any other industrial endeavour in a shareholder capitalist system, is to make money for shareholders' (Rogers, 2023). For arms dealers, a war that escalates, or develops into a violent stalemate, is a perfect war, because in both cases an insatiable demand for arms results, while each side is trying to improve its weaponry and tactics. The profit from such wars is placed above the value of human lives. As Ukraine continues to receive weapons and ammunition from NATO members, the war is becoming long, perfect for arms companies to profit from.

According to the Stockholm International Peace Research Institute, the USA is the world's largest arms exporter responsible for 38.6 per cent of international arms sales between 2017 and 2021, up from 32.2 per cent between 2012 and 2016. It has supplied arms to more than 100 countries in most of South America, Africa, the Middle East, Europe and Australia (Buchholz, 2022). The world's leading exporter of combat drones, however, is China. From Saudi Arabia to Myanmar, from Iraq to Ethiopia, Rasheed writes,

> governments and militaries across the globe are stockpiling Chinese combat drones and deploying them on the battlefield. In Yemen, a Saudi-led coalition has dispatched the Chinese aircraft, also known as uncrewed aerial vehicles or UAVs, as part of a devastating air campaign that has killed more than 8,000 Yemeni civilians in the past eight years.
>
> (2023)

During the 20th century, the trade in arms made viable conflicts that led to the killing of 231 million people. The first two decades of the 21st century have already seen short, long and perpetual wars, conventional, irregular, remote and proxy, at significant – and rising – human cost. Rather than committing to universal human rights, equality and justice, the pattern continues to be the production of more deadly weapons, the unscrupulous taking of life and the profiting from the suffering of others. As war is becoming more automated and robotic, there are more and more opportunities to lose control of the violence and the physical and mental agonies it unleashes.

On 3 November 2023, Russia launched a massive drone attack, hitting critical infrastructure in the west and south of Ukraine and destroying private houses and commercial buildings in Kharkiv.

> The air force said it shot down 24 'Shahed' drones out of 40 launched by Russia, the biggest drone attack in weeks to target Kharkiv in the northeast, Odesa and Kherson in the south and the region of Lviv on Ukraine's border with Poland in the west. Officials say Ukraine is bracing for a second winter of Russian air strikes on the energy system, which they warn is more vulnerable than it was last year as it has less excess capacity and little in the way of spare equipment.
>
> (Harmash, 2023)

Oleh Synehubov, Kharkiv's governor, said drones had hit civilian infrastructure and caused fires in and near the city of Kharkiv. He said eight people, including two children, required medical help due to acute stress. The governor of Lviv, Maksym Kozytskiy, stated that an infrastructure facility had been hit five times during attacks on his region, while Oleh Kiper, Odesa's regional governor, reported a strike on an infrastructure facility in the southern region.

On 5 October 2023, a Russian rocket blast turned a village cafe and store in eastern Ukraine into rubble, killing at least 51 civilians in one of the deadliest attacks in the war, according to President Zelensky and other top officials in Kyiv.

> Rescuers searched for survivors in the remains of the only cafe in the village of Hroza. Body parts were strewn across a nearby children's playground that was severely damaged by the strike. Cellphones were collected and put in a court-yard nearby, waiting to be claimed. Occasionally, one of them rang, lighting up a shattered screen. Around 60 people, including children, were attending a wake at the cafe when the missile hit, Ukrainian officials said.
>
> (Arhirova, 2023)

Although for more than a year Ukraine's military only killed Russian combatants, that began to change in the summer of 2023, when reports started to come in that civilians were getting killed in both countries. On 23 August, it was reported that civilians had been killed in attacks in both Russia and Ukraine, while Moscow was hit by drones for the sixth night in a row. Kyiv regularly hits Moscow and other cities deep inside Russia with drone attacks. In the Belgorod region on the border with Ukraine, according to authorities, three civilians were killed by Kyiv's forces. 'The Ukrainian forces launched an explosive device through a drone when people were on the street', Belgorod governor Vyacheslav Gladkov said on Telegram. For months, the region was targeted by shelling, drone attacks and cross-border incursions. Russian artillery hit two villages near the eastern Ukrainian city of Lyman, killing three people and wounding two others. Kyiv said in the early hours of Wednesday on 23 August that three people were killed in Torske, a small village in the east retaken by the Ukrainian army from Russian forces in October 2022. According to the Donetsk prosecutor's office, the victims were two women and a man – aged 63–88 – who were seated on a bench when shells hit. Two teachers were killed when Russia struck a village school in the northeastern Sumy region. 'In Romny, Sumy Region, the Russians destroyed a school and killed at least two educators', Igor Klymenko, Ukraine's interior minister, said on Telegram. They were still searching for two more of the school's employees believed to be under the rubble, he added. 'Three more people were injured', Klymenko said (Kolesnikova, 2023).

Ukraine's Sumy region has increasingly been targeted by remote attacks, while both Russia and Ukraine have also ramped up attacks in the Black Sea since the July collapse of a UN-brokered deal aimed at ensuring safe navigation for civilian grain shipments from Ukraine ports. Russian strikes on Ukraine's sea and river ports destroyed 270,000 tonnes of grain in the space of a month.

> "In total, 270,000 tonnes of grain have been destroyed in a month of attacks on ports," Infrastructure Minister Oleksandr Kubrakov said on social media. Earlier, it reported a Russian drone hit and damaged grain infrastructure in the southern Odesa region.
>
> (Kolesnikova, 2023)

The US-NATO factor in this war has highlighted the responsibility of Ukrainian leaders, as well as challenged the argument that this was a war of self-defence on the part of Ukraine, whose actions over two decades could be interpreted as causes of war. So keen was Ukraine to join NATO, a military alliance created precisely to counter the Soviet Union (and Russia), that in 2003 it sent forces to the newly invaded and occupied Iraq to fight alongside the invading/occupying forces: the USA and the UK. According to Iraq Body Count data collected in 2004, Ukrainian troops killed 13 Iraqi civilians as they shot at protesters in January and April.

Unlike the UN, NATO is not a universal organisation but a military and political alliance. Led by the USA, NATO has launched sustained military operations in the Balkans, in Libya and in Afghanistan. As for the USA, it has invaded and intervened militarily in several more countries, such as Iraq, Pakistan, Somalia and Syria. No matter how often it is claimed that NATO is purely a defensive alliance, 'its immense offensive military power projection capabilities that have been used in past wars say otherwise' (Puri, 2022: 87). NATO expansion is in line with US interests, it is not dictated or influenced by ethics, altruism or humanitarian considerations. Ukraine's actions are similarly in line with its own strategic interests and not guided by ethics, as the case of Iraq showed. If Ukraine joined the alliance, NATO's soldiers (Ukrainian or otherwise) would be stationed in Kharkiv, in Luhansk, in Donetsk, in Mariupol.

The involvement of NATO members, especially the USA and the UK, brings us back to the ethics of the War on Terror, where great/hegemonic powers fight wars of imperialism guided by self-interest, profit and the desire for power. What this war has made clear is the proliferation of drones in the utterly immoral and profit-driven global arms trade. With American, Iranian, British and Turkish-made weapons spreading death and destruction across Ukraine, the world is reminded of the dangers of unmanned aerial systems (UAS). From 2021 to 2022, drone sales went up to 57 per cent, an increase that has made it difficult 'to effectively ensure every echelon down to the operators of counter-drone systems is on the same page when it comes to strategic vision, operational mission, and tactical employment' (Pacheco, 2022); it has also made it difficult to keep up with the legal and moral implications, in this fast-paced evolution of warfare.

More than ten states have so far conducted drone strikes: the USA, Israel, the UK, Pakistan, Iraq, Nigeria, Iran, Turkey, Azerbaijan, Russia and the United Arab Emirates, while many other countries, including Saudi Arabia, India and China, maintain armed drones in their arsenals. All these states are navigating their own rules in the air, as there does not exist an international agreed-upon set of regulations. In the last few years, Serbia, Germany, Indonesia, Singapore, Algeria, Qatar and the Netherlands have acquired armed drones. Israel exports to 56 states, including Ukraine and Russia. The USA exports to 55, Ukraine included, but interestingly, China also exports armed drones to some of the same states, such as Ukraine and Russia.

Non-state actors all over the world are also using armed drones. Palestinian Islamic Jihad, for example, has used drones against Israeli forces. Venezuelan military defectors attacked President Maduro with drones armed with explosives during a speech commemorating the Bolivarian National Guard's 81st anniversary at a military parade in 2018. In Nigeria, Boko Haram fighters now 'have more sophisticated drones than the military' (Searcey, 2021). The militant group Harakat Tahrir al-Sham, with ties to al-Qaeda, attacked a Russian military base in 2018 using a 'swarm' of armed drones rigged with explosives. The Kurdish terrorist group PKK carries out drone strikes against Turkish soldiers, while the Libyan National Army (LNA) has conducted drone strikes against Libyan government forces since 2017, using Chinese-made Wing Loong II drones supplied by the United Arab Emirates. In Syria, at least four groups have used drones: Faylaq al-Sham (a Sunni rebel group fighting pro-Assad forces), the Turkistan Islamic Party (allied with al-Qaeda), Saraya al-Khorasani (Iranian-backed Iraqi Shia militia fighting ISIS forces) and Suqour al-Sham Brigades (part of the National Liberation Front). In the Philippines, the pro-ISIS Maute militants used drones against government forces during the Marawi battle. In Mexico, an armed drone was discovered in the arsenal of *Cártel de Jalisco Nueva Generación* (CJNG), an organised crime syndicate, in 2017. ISIS has trafficked commercial drone technology from at least

16 companies across at least seven countries, attaching munitions to quadcopters and small fixed-wing drones. The 'kamikaze' quadcopter drones had munitions strapped to them that could independently be released (Frew, 2018). In eastern Ukraine, the separatist group Donetsk People's Republic has used armed drones supplied by Russia since 2015.

Drones can be launched in mass and swarm attacks. A massed attack 'resembles several birds with decentralized flight patterns picking and choosing different prey while the second resembles an organized flock of birds converging on a single target' (Pacheco, 2022). Swarm attacks are particularly lethal, as they feature coordinated command and control, with 'one operator using an algorithm or tactical operations centre' (Pacheco, 2022). Multiple drones can communicate with each other remotely, and like a swarm of bees, they form a deadly and autonomous aerial army ripe for accidents. States are more likely to practise such attacks, which involve large numbers of drones, and China and Russia have focused on developing or using small UAS used in swarm attacks.

The war in Ukraine has highlighted a growing global arms trade that involves weapons with yet unclear ethics, legality and questions around accountability in a global war business where states profit from the development and sale of weapons to a variety of other states and to non-state actors, which use them for their own purposes. At the same time, the exporting states also use this technology, either domestically or abroad, sometimes in the client states. The example of the USA and Saudi Arabia shows this problem of moral and legal accountability. In 2014, the Saudi military ordered two CH-4 and five Wing Loong armed drones from China. By 2016, two Wing Loong drones had been deployed to the Najran Province, near the border with Yemen, where soon one of them was seen downed inside Yemen. The following year, Saudi Arabia ordered 300 Wing Loongs from China, and it was reported that a manufacturing plant would be opened in Saudi Arabia to facilitate this sale.

> Saudi Arabia has used its armed drones to carry out strikes in Yemen… In addition, US drones based in Saudi Arabia may be among those launching strikes into Yemen, further complicating attribution. Working out which air force has carried out what strikes is one of the inherent problems of drone technology and is likely to cause further accountability problems in the future.
>
> (Frew, 2018: 19–20)

The arrival of artificial intelligence and the development of sophisticated unmanned aerial vehicles have made drones more lethal and even more difficult to track, to investigate and to legislate. Drone attacks can come out of the blue, with no restrictions or consequences, because there are no rules guiding drone transfers, exports, imports and usage. Meanwhile, drone technology continues to evolve and, after the current war in Ukraine is over, what will likely follow is grey zone conflict: distant, proxy, dark, profit and interest-driven, and through delegation. And war, more than ever before, becomes an amoral or immoral business, most incompatible with just war principles.

As for Ukraine, it has 'tragically become a battle lab' for war technology, according to Ben Wallace, Secretary of State of the United Kingdom, but hopefully, the lessons learnt from it will inform the future. 'New technologies are not gimmicks, they're fundamentally key to how we fight a modern war', he declared a year and a half into the war, adding that analysing the strategies playing out in Ukraine would help make sure that the UK would be 'match fit for any future conflict' (Adams and Hancock, 2023). One of the lessons learnt is the power and use of electronic warfare, which is going 'up the priority list'. Once again, it is the realist's strategy, rather than any moral considerations, that takes priority and drives the action, the aim of which is to win at any cost, using the latest weapons with little to no control or regulation and with no accountability.

International justice

The ICC investigates and tries individuals charged with the gravest crimes: genocide, war crimes, crimes against humanity and the crime of aggression. The Court contributes to the global fight to end impunity, and through international criminal justice, aims to hold those responsible accountable for their crimes. A look at ICC cases reveals a picture of entirely dark-skinned leaders, primarily African.

*Al Hassan Ag Abdoul Aziz, Mali. Alleged member of Ansar Eddine and de facto chief of Islamic police.
Charges: suspected of crimes against humanity allegedly committed in Timbuktu, Mali, in the context of a widespread and systematic attack by armed groups Ansar Eddine/al-Qaeda in the Islamic Maghreb against the civilian population of Timbuktu and its region, between 1 April 2012 and 28 January 2013: torture, rape, sexual slavery, other inhumane acts, including, inter alia, forced marriages, persecution; and of war crimes allegedly committed in Timbuktu, Mali, in the context of an armed conflict not of an international nature occurring in the same period between April 2012 and January 2013: torture, cruel treatment, outrages upon personal dignity, passing of sentences without previous judgement pronounced by a regularly constituted court affording all judicial guarantees which are generally recognised as indispensable, intentionally directing attacks against buildings dedicated to religion and historic monuments, rape and sexual slavery.

*Omar Hassan Ahmad Al Bashir, Sudan. President of the Republic of Sudan since 16 October 1993 at the time of warrants.
Charges: five counts of crimes against humanity: murder, extermination, forcible transfer, torture and rape; two counts of war crimes: intentionally directing attacks against a civilian population as such or against individual civilians not taking part in hostilities, and pillaging; three counts of genocide: by killing, by causing serious bodily or mental harm and by deliberately inflicting on each target group conditions of life calculated to bring about the group's physical destruction, allegedly committed at least between 2003 and 2008 in Darfur.

*Ahmad Al Faqi Al Mahdi, Mali. An alleged member of Ansar Eddine, a movement associated with al-Qaeda in the Islamic Maghreb, head of the 'Hisbah' until September 2012, and associated with the work of the Islamic Court of Timbuktu.

Charges: found guilty as a co-perpetrator of the war crime consisting in intentionally directing attacks against religious and historic buildings in Timbuktu, Mali, in June and July 2012.

*Mahmoud Mustafa Busayf Al-Werfalli, Libya.
Charges: Mr. Al-Werfalli is alleged to have directly committed and to have ordered the commission of murder as a war crime in the context of seven incidents, involving 33 persons, which took place from on or before 3 June 2016 until on or about 17 July 2017 in Benghazi or surrounding areas, in Libya as well as Murder as a war crime in the context of an eighth incident which took place on 24 January 2018, when Mr. Al-Werfalli allegedly shot dead ten persons in front of the Bi'at al-Radwan Mosque in Benghazi, Libya.

*Saif Al-Islam Gaddafi, Libya.
Charges: Pre-Trial Chamber I considers that there are reasonable grounds to believe that, under article 25(3)(a) of the Rome Statute, Saif Al-Islam Gaddafi is criminally responsible as indirect co-perpetrator for two counts of crimes against humanity: • Murder, within the meaning of article 7(1)(a) of the Statute; and • Persecution, within the meaning of article 7(1)(h) of the Statute.

*Ali Muhammad Ali Abd-Al-Rahman, Sudan.
Abd-Al-Rahman is suspected of 31 counts of war crimes and crimes against humanity allegedly committed between August 2003 and at least April 2004 in Darfur.

Alfred Yekatom, Central African Republic. Former caporal-chef in the Forces Armées Centrafricaines and a member of parliament in the CAR.*
Yekatom is alleged to be responsible for crimes committed in this context in various locations in the CAR, including Bangui and the Lobaye Prefecture, between 5 December 2013 and August 2014. Crimes against humanity: murder, deportation or forcible transfer of population, imprisonment or other severe deprivation of physical liberty, torture, persecution and other inhumane acts. War crimes: murder, torture and cruel treatment, mutilation, intentional attack against the civilian population, intentional attack against buildings dedicated to religion, enlistment of children under the age of 15 years and their use to participate actively in hostilities, displacement of the civilian population and destruction of the adversary's property. This list is long and missing from it are British and American individuals, tried either for recent crimes committed in the Middle East, or retroactively.

As for justice in Ukraine, an International Criminal Tribunal for the Russian Federation, an *ad hoc* international criminal tribunal aimed at prosecuting the Russian Federation and senior Russian and Belarusian leaders for the Russian invasions of Ukraine as crimes of aggression, was proposed as a complement to the existing ICC investigation in Ukraine, as soon as Russia invaded. Several international bodies announced their support, including the Council of Europe, the European Commission, the NATO Parliamentary Assembly and the European Parliament. In April 2022, the Parliamentary Assembly of the Council of Europe (PACE) called for an ad hoc international criminal tribunal. In September 2022, the Council of Europe proposed to create a tribunal that would have a mandate to investigate and prosecute the crime of aggression committed by the political and military leadership of the Russian Federation, under customary international law. On 3 July 2023, it was announced that, the International Centre for the Prosecution of the Crime of Aggression against Ukraine (ICPA) had started its operations in the Hague, hosted by the European Union Agency for Criminal Justice Cooperation (Eurojust). The newly established Centre would be key to investigate Russia's crime of aggression against Ukraine and facilitate case building for future trials, providing 'a structure to support and enhance ongoing and future investigations into the crime of aggression and contribute to the exchange and analysis of evidence gathered since the start of the Russian aggression' (European Commission, 2023).

International justice seemed to be served a few years earlier, at the International Criminal Tribunal Yugoslavia, established to try Serb and Croat leaders accused of committing crimes against civilians in Bosnia Herzegovina, Croatia and Kosovo, from 1991 to 1999, leaders like Slobodan Milošević. In her opening remarks in the trial of Milošević, Prosecutor Carla Del Ponte declared: 'This Tribunal, and this trial in particular, give the most powerful demonstration that no one is above the law or beyond the reach of international justice' (United Nations International Criminal Tribunal for the Former Yugoslavia, Slobodan Milošević Trial).

When the Tribunal indicted Milošević in May 1999 for crimes in Kosovo, he was the first sitting head of state to be charged with war crimes by an international tribunal. The Tribunal is the first international law enforcement agency to respond to war crimes in real time. It is claimed that the proceedings against Milošević were a turning point for international justice because the Tribunal proved that not even the president of a country could use his position of high office to claim immunity from prosecution and that societies need not live for decades in denial about the role their highest leaders played in committing crimes. 'Indeed, the multitude of evidence submitted in the Milošević trial, much of it never seen before, may be considered its greatest achievement: it remains a part of the public record … a barrier against malign attempts to revise history' (United Nations International Criminal Tribunal for the Former Yugoslavia, Slobodan Milošević Trial).

The trial opened on 12 February 2002, by which time the Tribunal had confirmed indictments against Milošević for crimes in Croatia and Bosnia Herzegovina, following years of painstaking investigations into many dozens of crimes, the direct perpetrators who committed them, the mid-level leaders they reported to and finally the highest-level authorities with whom the Prosecution alleged Milošević had conspired.

Over 20 years after the opening of Milošević's trial, in November 2023, Del Ponte (chief prosecutor of the International Criminal Tribunal for the Former Yugoslavia 1999–2007 and the International Criminal Tribunal for Rwanda 1999–2003) and Graham Blewitt (deputy prosecutor of the International Criminal Tribunal for the Former Yugoslavia 1994–2004) commented on the importance of prosecuting war crimes, which they termed a vital task. As former international war crimes prosecutors, they recognised how important it was to take steps to provide accountability for victims of atrocities. International criminal law plays a pre-eminent role in combating impunity and contributing to the deterrence of further crimes. In relation to Palestine, they noted, the ICC's potential was yet to be realised. Indeed, international prosecutors face considerable legal, practical and political challenges in fulfilling their mandates.

> There's an inevitable selectivity at play when prosecutors choose cases, even when tasked with addressing international crimes occurring in a particular territory and over a defined time period. Nevertheless, in the context of the vicious ethnic conflicts in former Yugoslavia and the genocide in Rwanda, we wouldn't have fulfilled our solemn duty by only prosecuting softer targets. And given the ICC's global reach, its prosecutor has a far broader jurisdiction to contend with. The court's jurisdiction extends to the territory of over 120 state parties as well as the nationals of those countries, and even further if the U.N. Security Council calls upon the court to act — as it did in Darfur and Libya. Thus, (chief prosecutor) Khan is engaged in a constant balancing act of competing demands, limited resources and varying degrees of state cooperation.
>
> (Del Ponte and Blewitt, 2023)

Prosecutor Karim Khan has emphasised the importance of accountability and of combatting impunity. In relation to Russia's aggression against Ukraine, he has provided tremendous support to his Ukrainian counterparts. His indictment of President Putin is without precedent for the sitting head of state of a permanent UN Security Council member to be subject to such an arrest warrant. And while visiting the Rafah Crossing in October 2023, Khan emphasised the need to ensure that 'law is on the frontlines' in Gaza (Samuels, 2023). Strong words by Khan, Del Ponte and Blewitt. Yet it was in March 2021 that ICC prosecutor Fatou Bensouda stated that he would be confirming the opening of an investigation by the Office of the Prosecutor of the ICC the situation in Palestine. The investigation, he stated, would focus on 'crimes within the jurisdiction of the Court allegedly committed in connection with this situation since 13 June 2014, the date to which the referral of the Situation in Palestine to my Office refers' (International Criminal Court, 2021).

Given the speed at which tribunals to try crimes committed by Russians, Serbs and Croats were set up, it is (perhaps) surprising that two-and-a-half years after Bensouda's confirmation of the opening of an investigation into the situation in Palestine, not only had nothing happened (no trials, no arrest warrants) but, in addition, Israeli PM Netanyahu was secure enough in the knowledge that Israel could kill with impunity to launch attacks on Palestinian civilians that, two months in, have surpassed even the deaths of Iraqis in Shock and Awe. According to Iraq Body Count figures, 991 Iraqi civilians were killed in the first three days of Shock and Awe, 2,314 during the following week, 2,315 the week after that and 1,299 from day 18 to day 24. In Gaza, Israeli air strikes killed 580 civilians in the first three days, 2,440 during the following week, 2,791 the week after that and 2,817 from day 18 to day 24, based on figures provided by the Ministry of Health in Gaza (Iraq Body Count, Twitter, 31 October 2023).

Israel is able to kill without any accountability because it is America's biggest ally in the Middle East, and alliances shape both regional and global security dynamics, as well as determine questions on justice. The USA was the first country to offer de facto recognition to the new Israeli government in 1948 when the Jewish state declared independence. Seventy-six years later, Washington has long been Israel's strongest military and diplomatic ally. As Israel declared itself 'at war' on 7 October, US officials reiterated their unwavering support: diplomatically, financially and militarily. US President Biden was unequivocal in an address at the White House, declaring, 'We stand with Israel … And we will make sure Israel has what it needs to take care of its citizens, defend itself, and respond to this attack' (Baker, 2023).

Appearing alongside Netanyahu on 12 October, as Israel was about to embark on its retaliatory campaign that included air strikes and ground operations that would take thousands of innocent lives and annihilate entire families, US Secretary of State Antony Blinken said, 'You may be strong enough on your own to defend yourself, but as long as America exists, you will never ever have to. We will always be there by your side' (Narea, 2023). The UK too expressed its continued support of Israel, even as it acknowledged the scale of death and destruction it was causing. On 8 December, two months into the war, British Ambassador Barbara Woodward said in her statement at the UN Security Council meeting on Gaza:

> The terrible and heart-wrenching suffering of innocent Palestinians, including many women and young children, is a humanitarian tragedy unfolding before our eyes (…) The sheer scale of civilians killed is shocking and the fact that 80% of the population has been displaced in Gaza cannot continue. The UK continues to support Israel's right to defend itself against Hamas terrorism as it seeks the return of over 100 hostages who are still held in Gaza. But we are absolutely clear that Israel must be targeted and precise in achieving that goal.
>
> (Woodward, 2023)

No calls for arrests, tribunals or any other actions to be taken to hold anyone accountable for the thousands already killed. Only a recommendation that Israel needs to be more careful.

As for French President Macron, while he had strong words for Israel, his main message was to 'avoid escalation'. Calling for a ceasefire, humanitarian pauses and increased aid to Gaza (Varma and Huggard, 2023), he too failed to speak of war crimes and justice. Four months into the war, 30,642 Palestinian civilians had been killed, including 13,000 children and babies; over 67,000 had been injured and 2,000,000 became displaced. Hundreds of thousands of homes were destroyed, 334 schools and 235 healthcare facilities were damaged (Euro-Med Human Rights Monitor, 2024).

In the case of Syria, not a Western but a Russian ally that the USA, the UK and France have been bombing on and off for a decade, on 15 November 2023, French criminal investigative judges issued arrest warrants for President Bashar al-Assad, his brother Maher al-Assad and two other senior officials – General Ghassan

Abbas, Director of Branch 450 of the Syrian Scientific Studies and Research Center, and General Bassam al-Hassan, Presidential Advisor for Strategic Affairs and liaison officer between the Presidential Palace and the SSRC – over the use of banned chemical weapons against civilians in the town of Douma and the district of Eastern Ghouta in August 2013, in attacks which are said to have killed over 1,000 people These arrest warrants refer to the legal qualifications of complicity in crimes against humanity and war crimes. According to Mazen Darwish, founder and director general of the Syrian Center for Media and Freedom of Expression (SCM),

> The French judiciary's issuance of arrest warrants against the head of state, Bashar al-Assad, and his associates constitutes a historic judicial precedent. It is a new victory for the victims, their families, and the survivors and a step on the path to justice and sustainable peace in Syria.
>
> (SCM, 2023)

He added, 'No one is immune. And we expect the French authorities to respect the victims' suffering and rights along with the decision of the French judiciary' (SCM, 2023). The judicial action by the French judges comes after a long criminal investigation into two chemical weapons attacks in August 2013 by the Specialized Unit for Crimes against Humanity and War Crimes of the Paris Judicial Court. The investigation was opened in response to a criminal complaint based on the testimony of survivors, filed in March 2021 by the Syrian Center for Media and Freedom of Expression (SCM) and by Syrian victims, and supported by hundreds of items of documentary evidence, including photos and videos. With these arrest warrants,

> France is taking a firm stand that the horrific crimes that happened ten years ago cannot and will not be left unaccounted for. We see France, and hopefully, other countries soon, taking the strong evidence that we have gathered over years and finally demanding criminal responsibility from the highest-level officials,' said Hadi al Khatib, founder of Syrian Archive and Managing Director of the project's host organisation, Mnemonic.
>
> (SCM, 2023)

Steve Kostas, Senior Managing Lawyer at the Open Society Justice Initiative, noted that this was the first time a sitting head of state had been the subject of an arrest warrant in another country for war crimes and crimes against humanity. He declared it a 'historic moment', because, with this case, France has an opportunity to demonstrate and establish the principle that there is no immunity for the most serious international crimes, even at the highest level. Aida Samani, Senior Legal Adviser at Civil Rights Defenders, expressed the hope that the arrest warrants would send a message 'loud and clear' to the survivors, and everyone affected by atrocity crimes in Syria, that 'the world has not forgotten them and that the fight for justice will continue' said (SCM, 2023).

In other Syrian cases, the Paris court's Specialized Unit for Crimes against Humanity and War Crimes has previously issued arrest warrants for seven other senior officials in the Syrian regime, including the current head of Syria's National Security Bureau, Ali Mamlouk. In October 2020, the Open Society Justice Initiative, Syrian Archive and SCM filed a similar complaint before the Office of the German Federal Public Prosecutor on the Syrian government's sarin attacks on al Ghouta in 2013 and on Khan Shaykhun in 2017.

The principle of extraterritorial jurisdiction can be used by courts to investigate and prosecute international atrocity crimes committed on foreign territory under certain circumstances. While prosecuting crimes committed by the Assad regime is important and admirable, what is missing from these investigations, arrests and prosecutions regarding war crimes, is, again, any serious consideration of crimes committed by Western states, France included. In Syria alone, the USA, the UK and France (alongside other NATO member states) have been involved in a bombing campaign since 2014. Airwars found that between 2014 and 2018, up to 9,600 Syrian civilians were killed in coalition actions. Where is the extraterritorial jurisdiction, when it comes to those killings? And where is it, when it comes to the killings of another NATO member and Western ally, Turkey? Iraq Body Count in collaboration with Community Peacemaker Teams documented Kurdish victims of Turkish air strikes in Iraq. The collaboration led to 'Civilian Casualties of Turkish Military Operations in Northern Iraq (2015–2021)', which published these key findings:

The Turkish Armed Forces have conducted at least 88 cross-border aerial, artillery, and ground attacks which caused civilian deaths and injuries within the borders of Iraq between 1 August 2015 and 31 December 2021. Despite the Turkish Government's claims of solely targeting the insurgents of the Kurdistan Workers' Party (PKK) and their affiliated groups, the eighty eight attacks by the Turkish Armed Forces have caused the death of 98 to 123 civilians and non-belligerents of the conflict and injury of 134 to 161 civilians and non-combatants.

(End Cross-border Bombing, 2022)

In 2022, a new military operation was launched by the Turkish Armed Forces, codenamed Claw-Lock, within the territory of Iraqi Kurdistan. The operation consists of large-scale aerial bombardments, as well as the deployment of special forces on mountain ranges in elevated areas up to 12–15 km south of the Turkish border. As for Turkish incursions into Syria, they have been fraught with human rights abuses. In 2019 and 2020, Turkey and the Syrian National Army (SNA), a non-state armed group backed by Turkey in northeast Syria, indiscriminately shelled civilian structures and systematically pillaged private property held by the local Kurdish population, arrested hundreds of people and summarily killed Kurdish forces, political activists and emergency responders in areas they occupy in northeast Syria. According to the UN Commission of Inquiry on Syria, Turkish-backed forces committed at least 30 incidents of rape. Turkey's 2018 military offensive in Afrin resulted in the deaths of dozens of civilians and displaced tens of thousands, according to the UN. Human Rights Watch investigated three attacks in northwest Syria that claimed the lives of 23 civilians. According to local activists, at least 86 incidents of abuse appeared to amount to unlawful arrests, torture and disappearances (Human Rights Watch, 2022).

With Israel's attacks on Gaza and the West Bank, which have so far killed over 11,000 civilians (1 every 200 people), the Middle East – even more than Ukraine – is on a deadly precipice. Yet arrest warrants are not forthcoming for Western states and Western allies, and neither is justice for victims.

> For victims of international crimes, their faith in the international criminal justice system has been sorely tested. There's a perception that there are double standards at play, and that the ICC disproportionately focuses on African countries and non-state actors, while allowing Western countries and their allies to evade their responsibilities.
>
> (Del Ponte and Blewitt, 2023)

Global security dynamics and their impact

As the war in Ukraine continues, to preserve information and evidence regarding war crimes, investigative journalists and human rights activists have created an online register of war criminals to document incidents, which already includes over 180,000 Russians. Journalists of the Ukrainian independent investigative agency *Slidstvo.Info* and activists of the Anti-Corruption Headquarters have already identified 182,000 Russian war criminals involved in the war against Ukraine. The Register of Russian War Criminals was created in March 2022 and information has been input since then. The Register of Russian War Criminals website begins with a section where one can search for Russian soldiers already on the database. They can also see a list of soldiers already entered. Each soldier has his own page. By clicking the name, the information is accessed (Register of Russian War Criminals).

Serhiy Mytkalyk, co-founder of the Register of Russian War Criminals, said that in addition to using journalistic investigations and Russian telegram channels, to identify or locate Russian soldiers involved in the incidents, they also

> cooperate with investigative authorities, including the Prosecutor General's Office. However, for objective reasons, the pace of investigation of these crimes is not as fast as [the public] expects. As of today, we have about 170 verdicts against Russians, including State Duma deputies.
>
> (Zakharchenko, 2023)

Ukraine is the first country to investigate war crimes during active hostilities. So far, there are '109,538 proceedings under the article "violation of the laws and customs of war". There are 450 suspects in these crimes, indictments have been drawn up against 291 people, and guilty verdicts have been handed down against 66 people', Oleksandr Zyuz, head of the Department for Intergovernmental, State and Non-Governmental Organizations of the War Department of the Prosecutor

General's Office, said at a press conference (Zakharchenko, 2023). The ultimate goal is to bring to justice not only every Russian soldier involved in the war against Ukraine but also the Russian leadership.

Victims or witnesses of Russian war crimes can report incidents and provide evidence to the Office of Prosecutor General at https://warcrimes.gov.ua/, available in both Ukrainian and English. 'We store information for the administration of justice', the website says.

Indeed, justice – and the need for it – is on everyone's lips in the Western world, when it comes to Russia. After all, Russia poses the 'most immediate threat' to UK security, according to ex MI6 chief Sir John Sawers (Soteriou, 2022). Sir John Sawers – who led the intelligence service between 2009 and 2014 – declared Russia the larger, immediate threat and China a long-term threat. A year earlier, in 2021, British PM Johnson had announced that Russia was Britain's top security challenge, while also affirming the security challenges presented by China. Speaking in parliament after the release of the 116-page report 'Global Britain in a Competitive Age', Johnson criticised China for its mass detention of Uyghur people in Xinjiang province and treatment of democracy advocates in Hong Kong. But the report reserved its most significant criticisms for Russia, which it regards as the most acute direct threat to the UK. 'Russia is the most acute threat in the region and we will work with NATO Allies to ensure a united Western response, combining military, intelligence and diplomatic efforts', the report says, pledging to work within the alliance to 'deter nuclear, conventional and hybrid threats to our security, particularly from Russia', while drawing attention to 'geopolitical and geoeconomic shifts, such as China's increasing power and assertiveness internationally, the growing importance of the Indo-Pacific to global prosperity and security, and the emergence of new markets' (Policy Paper, 2021).

The document mentions:

- The growing threat from authoritarian states that 'export their domestic models, undermine open societies and economies, and shape global governance in line with their values'.
- The specifically digital dimension of this challenge.
- The use of illicit finance to undermine security.
- Disinformation, cyber-attacks and election interference.
- Espionage.

It concludes that Russia presents the full spectrum of threats to the UK and its allies.

The East–West Cold War, or New Cold War, defines, identifies and connects old and new threats. It calls them 'global', but what they really are is Eastern threats to the West: the threats of (1) Russia and China and the threats from (2) the Middle East (the War on Terror). For the majority of the time, in the 21st century, those threats are routinely met or responded to in the war tradition, where military responses to political problems are the state's default position. Those military responses necessarily come with human casualties (of civilians and combatants), war crimes, destruction, displacement and all kinds of trauma.

What is the answer? First, the rejection of war as a legitimate method of resolving disputes. From a Kantian perspective, some wars may be more legitimate than others, 'but none is fully recommended in a legal or moral sense' (Williams, 2012: 60). War impacts – transforms even – all humans involved in it: it does not only impact the victims but also the aggressors, as well as those fighting a war of self-defence. Let us remember Kant again: 'A state against which war is being waged is permitted to use any means of defence except those that would make its subjects unfit to be citizens' (Williams, 2012: 66). And the more technological war becomes, the less human the relations of those in it become. As the killing distance increases, 'so the moral perception of the act of killing diminishes: an adversary who is never seen can be killed with equanimity' (Coates, 2016: 102). The human security approach renounces war: close, distant, aggressive and defensive. It adopts a type of pacifism that rejects militarism in all its forms and all its expressions and replaces it with pacifist values, that is, values that protect and nurture life, rather than kill and destroy; values that glorify peace, rather than war.

War is seen as rooted in human nature by more than political realists. It then follows that war is inevitable and ineradicable. However, the propensity to go to war (discernible throughout man's history) can be seen as the product of a social and historical reality that is not unchangeable: 'not a static entity, but a social and historical phenomenon. War may be rooted in humans, but only in the sense of a human nature fashioned and produced by the existing war system' (Coates, 2016: 108). The answer that will save, spare and protect millions of human lives lies in transforming the structures of violence that induce and glorify violent behaviour at national and international levels to achieve the genuine pacification of society and the transformation of a military culture into a pacific one.

The world is committed to war and 'in countless ways, non-military institutions and practices serve military ends' (Holmes, 1989: 268). Over centuries of historical development, 'societies have been transformed into systems of war that are geared socially, economically and politically to the maintenance and often to the glorification of war. The "garrison state" is the historical norm rather than the exception' (Coates, 2016: 108).

Key economy sectors are involved with the defence and armament industries, which means that millions of people have a vested interest in the perpetuation and expansion of the business of war.

The education system too, formally and informally, cultivates a disposition to war. From the earliest age children – particularly male children – are imbued with military values. In both secular and religious circles language is dominated by a military idiom and imagery. In such varied and mutually reinforcing ways our social institutions secure a central place for war and ensure its widespread and ready acceptance. The roots of war lie deep within our culture.

(Coates, 2016: 108)

What is needed is the radical transformation of institutions, beliefs and practices in every state in the world, but especially in the more powerful states like the USA, the UK, Israel and Russia, where militarism is deeply and tightly embedded in society, where the narrative is one of moral armies, clean and just wars. In the 21st century, academics assert and reassert the importance of decolonisation, but of equal importance for global security and justice is demilitarisation. The connections between nations, nationalism and military valour, being 'ready to die' for the homeland and honouring one's heroic ancestors, can be seen in the national anthems that tell of a glorious past and a glorious present, filled with bravery, virtue and self-sacrifice, usually emerging through wars and shaping the nation's mission and destiny.

In Ukraine, the human cost resulting from Russia's war of aggression is already being addressed. Some of the perpetrators of war crimes are already being held accountable, much like those in the former Yugoslavia. In 1993, in Brussels and New York, diplomats were wrangling with a difficult choice: pursue a war crimes tribunal endorsed by the UN General Assembly, or forgo such a trial to cajole Serbia into a peace plan with the EU. 'Dropping the trial, the residents of Bosnia and Herzegovina told Neier, would be a betrayal. Justice for war crimes, ethnic cleansing, and genocide cannot be a "bargaining chip," he argued' (Ling, 2022). The International Criminal Tribunal for the Former Yugoslavia commenced later that year.

Ukraine's legal system does not have the capacity to handle the number of cases that are expected: then-Prosecutor General Iryna Venediktova told the BBC in 2022 that they had received reports of some 21,000 war crimes. A cohort of international law experts say that now is the time to lay out how the world intends to investigate and prosecute allegations of war crimes committed in Ukraine.

'As the conflict goes on, more and more Russian soldiers, colonels, and generals are going to fall into the custody of the Ukrainians,' Michael Scharf, co-dean of Case Western Reserve University's School of Law, told *Foreign Policy*. 'And as that happens, there's going to be a lot of interest in starting the prosecutions. … I cannot imagine that the current Ukrainian court system can handle that'.

(Ling, 2022)

A draft law written by Scharf and others contemplated a high war crimes court located in Ukraine and staffed by Ukrainian judges, with international observers, foreign advisors, a security guarantee provided by the West and considerable funding. Scharf pointed to a £2.5 million donation from the UK to fund some of the investigative work. The ICC will pursue high-ranking individuals. Scharf sees international bodies like the ICC operating in conjunction, not in competition, with the Ukrainian court. Any decisions about how to share war crimes cases would likely fall to the Ukrainian prosecutor general. The ICC could pursue those who won't be extradited to Ukraine, for example. 'People want to see this internationalization to avoid the appearance of bias', Stephen Rapp, the former US ambassador-at-large for war crimes issues, said at the Public International Law and Policy Group roundtable event in July (Ling, 2022).

There are benefits to pursuing these charges now, Scharf argues, pointing to the political pressure placed on Milosevic through his indictment, following which he agreed to the independence of Kosovo. When it comes to the Russian leadership, the indictments themselves could have a strategic benefit, as they erode their power, their popularity and their positions of authority.

For over a century, the international community has tried to install binding accountability measures. With each war or armed conflict, conventional or irregular, that infrastructure has started to take on a more consistent and predictable appearance. Regardless of whom Ukraine and the international community manage to arrest and prosecute, Ling maintains, 'fulfilling the promise made at Nuremberg will be a meaningful contribution to international law' (Ling, 2022).

However, the war need not have happened at all, and the hundreds of thousands killed on both sides need not have died. The biggest factor was the East–West dynamic of the New Cold War, which not only led to the war, which not only employed the securitisation of Russia and Russians, but also advanced and accelerated national and global appeals to justice. That same dynamic means that hundreds of thousands of others who have died terrible deaths – and are still dying – in the War on Terror (also part of the New Cold War) will never receive justice, and millions will continue to live in fear, economic, community and political insecurity, with little food or access to healthcare, displaced, in grief and in anger. As militarism rules and drones populate the skies over cities, the Register of American War Criminals, the Register of Israeli War Criminals, the Register of British War Criminals, the Register of Turkish War Criminals and the Register of French War Criminals remain empty of names.

Conclusion

New Cold War of the 21st century and human security

As we have argued in the previous chapters of this book, 21st-century international affairs have progressively turned out to be more and more violent, and respect for human lives, safety and security of civilians were sacrificed to achieve the military goals of the states and economic interests of multinational corporations. Even though early in this century there was an emphasised attempt by the UN, and several global institutions, that we need to recognise security differently and widely to put individual citizens' security at the centre of our understanding, to have human security as the central goal of international affairs, as we moved further in the century almost all global powers have tended to disregard human security more and more in the context of an increasing number of wars, civil wars and various calamities, all of which caused extreme distress to human life and living conditions. For those many violent conflicts, from the US-led War on Terror to Russia's violent military invasion of Ukraine, from Syrian civil war to the 2023 war in Gaza, of course there are specific local and regional reasons. However, at a general level, at the root of this sharply increased insecurity for the peoples of the world, there is a combining factor: changing the balance of power in the global inter-state system, or in other words the disequilibrium emerged as a result of fast-changing positions and sharply increased clash of interests among global superpowers.

Today's conflicts are a reflection of this fragmented international system that is deeply antagonistic, and therefore largely powerless. Many such conflicts have become proxy wars, fuelled by rival foreign powers such as the USA, Russia, France, Iran or Turkey. Whether it be the Russian–Ukrainian conflict or the one in Gaza between Israel and the Hamas movement, the problem seems to be the same: there is no effective international body capable of promoting or imposing a peaceful resolution and providing protection to the civilian population. The 'rules-based international order' seems like a toothless tiger – there is still too much reference to it, but in actual reality, it has no power. Therefore, talking about a 'rules-based international order' appears nothing more than a blatant PR trick.

Italian philosopher Antonio Gramsci wrote in his *Prison Notebooks* in 1930 that 'the crisis consists precisely in the fact that the old is dying and the new cannot be born. In this interregnum, a great variety of morbid symptoms appear'[1] (Gramsci, 1971: 276). In 1930, Gramsci was preoccupied with the breakdown and collapse

DOI: 10.4324/9781003414827-6

of the international order which was the dominant pattern in international relations after World War I. When struggling to understand the rise to power of Benito Mussolini in Italy, a similar possibility in Germany, and the serious unbalancing impact of these extreme rulers in the international system, Gramsci used the term, 'morbid phenomenon' or 'morbid symptom'. For Gramsci, Mussolini's rise to power was one such morbid symptom. The term 'interregnum' was originally used to denote a time lag separating the death of one royal sovereign from the enthronement of the successor. Interregnum here, as referred to by Gramsci, can be understood with a new wider meaning as a period where one arrangement of hegemony is waning, but prior to the full emergence of another. Likewise, it seems that in the third decade of the 21st century, our world is once again going through an interregnum. It is poised between inward-looking old hegemonic powers (the US and European states) and new emergent ones (China, Russia and other emerging powers). The chief FT columnist Martin Wolf was referring to exactly this in 2018 when saying that 'today, ...the liberal international order is sick' (Wolf, 2018). The US-centred world of post-World War II is losing its supremacy and is slowly and painfully being replaced by a new international system formed by the arrival of new actors. This is basically what causes the breakdown of the global order, and its norms, principles and international agreements are losing their authority in an increasingly anarchic global jungle of war and violence. The West, collectively, does not have the means to back up its policies in the Middle East, Africa, Ukraine and Southeast Asia.

The new emerging powers, on the other hand, aspire to a new world order of a multi-polar global system, but lack any real leadership capacity. They are not currently in a position to impose a stable authority and system of control upon various regional and global conflicts. Therefore, leadership, order and regional and global governance are no longer assured. The world is currently in a fragile imbalance as the global hegemon's decline continues, described aptly by Martin Wolf in 2017: 'US power has retreated both geopolitically and economically, and we are living, once again, in an era of strident nationalism, and xenophobia', and one can add this, extreme and uncontrolled violence of the jungle (Wolf, 2017). The 'New Cold War', the USA on one side and Russia and China on the other, is a direct product of this 'interregnum', breakdown of global order. It is the direct consequence of a process that Giovanni Arrighi described as a 'hegemonic transition' within a period of systemic chaos. This protracted period of hegemonic transition from the Euro-Atlantic core to Asian powers, especially China and India, like every other period of hegemonic transition and instability in which 'the old is dying and the new cannot be born', has created more morbid symptoms and uncontrolled violence (Arrighi and Silver, 2001).

The multi-level conflicts of the New Cold War of the 21st century, as a global security dynamic, led by the workings and requirements of military-industrial complexes, have already resulted in very large numbers of casualties all over the world. This large-scale and intense level of state and transnational violence are now seen not to be exceptional but the normal functioning of the system. The emphasis is, as in the 19th and early 20th century, on military power, state security, accumulating more and more weapons and gaining the upper hand in preparing for and sustaining wars. The result of this heightened security environment is a sharp increase in insecurity for ordinary citizens, the real losers of the wars and conflict – the unarmed population. It is almost that the inter-state global system needs to be in a constant state of emergency in order to function in an effective and profitable way.

A clutch of multinational arms manufacturers, often based in the USA, profit from these wars and conflicts. Business, in recent years, has been on the up. All these military conflicts also mean the destruction of infrastructure, which allows for contracts in reconstruction. Corporations such as Lockheed Martin, Boeing, Halliburton, Bechtel, Skylink-USA, Stevedoring Services of America, BearingPoint, General Dynamics and many others would, in all likelihood, collapse were it not for arms sales and reconstruction contracts. The governments who oversee such a trade present this deadly trade as champions of a non-existent 'rules-based international order'. Many of the biggest corporations in the world rely on such state contracts, and all these business links simply become a circular relationship between the states and multinational corporations. It would not be an exaggeration to claim that the military-industrial complex, which operates as an informal alliance between multinationals and the states, has become central to the global inter-state system in the 21st century. The public in many countries is told that these arms produced, sold and accumulated are for the national interest, which means, according to governments, they are for the protection of people against threats, but mostly non-existent threats. In reality, the production of more and more deadly weapons makes their usage more likely.

> Their [the world's arms dealers] primary function, like that of any other industrial endeavour in a shareholder capitalist system, is to make money for shareholders while ensuring decent salaries and even more decent bonuses for the CEO and senior colleagues. To them, a 'perfect war' is one that degenerates into a violent stalemate that creates an insatiable demand for arms and the replacement of worn-out equipment, while at the same time, each side constantly tries to improve its weaponry and tactics.
>
> (Rogers, 2023)

Just like any other product produced in the global economy, the weapons and bombs made have to be sold and used, otherwise the war business cannot continue. Simply because war is good business, with more and more wars and conflicts, the nations tend to buy more and more weapons, meaning more and more profitable business for the military-industrial complex. Whenever there is a war, or civil war, in the world, the military-industrial complex sells weapons to the warring sides and makes money, no matter who wins or loses.

The amount of money involved is enormous. During the War on Terror, for instance, all those corporate firms, arms and support services, logistics, weapons maintenance, military training and security achieved astronomic levels of profit, much more than any peacetimes.

Corporations large and small have been, by far, the largest beneficiaries of the post-9/11 surge in military spending. Since the start of the war in Afghanistan, Pentagon spending has totalled over $14 trillion, one-third to one-half of which went to defence contractors. Some of these corporations earned profits that are widely considered legitimate. Other profits were the consequence of questionable or corrupt business practices that amount to waste, fraud, abuse, price-gouging or profiteering.

(Bandow, 2021)

One of the most powerful instruments of the military-industrial complex to promote its business is its connections to foreign policy think tanks to shape the national and global agenda on security and defence. More than 75 per cent of the foreign policy think tanks in the USA are funded by defence corporations. Many such think tanks receive millions of dollars every year from such companies, in return they endorse/encourage articles and reports that are largely supportive of the conducts and interests of the military-industrial complex. Whenever there is media coverage on issues of war and peace, all too often researchers and affiliated analysts from these think tanks provide 'expert' opinions.

In the first two decades of the 21st century, many violent wars, and civil wars, have already induced a state of perpetual paranoia and justified the unjustifiable: black sites, unlawful detentions, 'waterboarding', 'walling', 'rectal feeding', targeted assassinations, unlawfully detaining suspects at sites abroad or sea and war crimes of all kinds. Dick Cheney, US Vice President, was not shy in justifying these unlawful practices:

We also have to work, though, sort of the dark side, if you will … if we're going to be successful. That's the world these folks operate in, and so it's going to be vital for the US to use any means at our disposal, basically, to achieve our objective.[2]

Millions of civilians have died and more displaced by wars and civil wars that have been made possible by this very profitable but deadly trade. This militarised accumulation of weapons has unleashed cycles of destruction and reconstruction, generating enormous profits for the ever-expanding military-industrial complex. In a report issued in May 2023, *the Cost of War Project* at Brown University estimated that at least 4.5 million people died as a result of the wars launched since the 9/11 attacks in 2001. This was based on the figures from the wars in Afghanistan, Iraq, Libya, Somalia, Syria and portions of Pakistan affected by the spillover of the war in Afghanistan.

The Project puts forward four interconnected causes for the large number of casualties as a consequence of these wars:

- Economic collapse, loss of livelihood and food insecurity.
- Destruction of public services and health infrastructure.
- Environmental damage.

• Mental trauma and violence.

There are many other outcomes of these conflicts, like unexploded ordnance in massive quantities, environmental degradation, widespread and long-standing mental health issues and the destruction of essential infrastructure important for public health and welfare, all of which will continue to haunt even the next generations in those countries.

> … more than 7.6 million children under five are suffering from acute malnutrition, or wasting, in Afghanistan, Iraq, Syria, Yemen, and Somalia. 'Wasting' means, simply, not getting enough food, literally wasting to skin and bones, putting these children at greater risk of death, including from infections that result from their weakened immune systems.
>
> (Savell, 2023: 11)

In the third decade of the 21st century, it is hard to be optimistic about the future. Our world has truly entered a period of great upheavals and uncertainties, of momentous changes, fraught with serious shocks, risks and fragilities. The scars of the current wars are deep and the impact will last for generations. There is a sharply rising spectre of worldwide instability, wars and civil wars. The future of current conflicts is far more uncertain, which might explode into something global and seriously more damaging. As the Russia–Ukraine war was getting more and more hopeless with mounting civilian casualties in Ukraine, the massacres that took place on 7 October 2023 in southern Israel started a much more deadly conflict. Hamas militants stormed through villages and kibbutzs in Israel, killing 1,200 people and taking 240 hostages. In response to Hamas's attack came the excessively violent Israeli response against a backdrop of an embargo on water, electricity and fuel, with no way out for the people of Gaza. The ferocity of the bombardments, with thousands of civilian casualties and 1.9 million out of 2.3 million of Gaza's inhabitants being displaced, quickly came to overshadow the nature of the 7 October events.

'Even war has rules', UN secretary-general António Guterres told Security Council, demanding all parties in the Middle East uphold International Humanitarian Law and unrestricted aid for Gaza.

> Nothing can justify the deliberate killing, injuring and kidnapping of civilians – or the launching of rockets against civilian targets. All hostages must be treated humanely and released immediately and without conditions. I respectfully note the presence among us of members of their families.
>
> It is important to also recognize the attacks by Hamas did not happen in a vacuum. The Palestinian people have been subjected to 56 years of suffocating occupation. They have seen their land steadily devoured by settlements and plagued by violence; their economy stifled; their people displaced and their homes demolished. Their hopes for a political solution to their plight have been vanishing.

But, the grievances of the Palestinian people cannot justify the appalling attacks by Hamas. And those appalling attacks cannot justify the collective punishment of the Palestinian people.

Even war has rules. We must demand that all parties uphold and respect their obligations under international humanitarian law; take constant care in the conduct of military operations to spare civilians; and respect and protect hospitals and respect the inviolability of UN facilities which today are sheltering more than 600,000 Palestinians.

(UN, 2023)

With the death toll passing 20,000 in the Palestinian territories, South Africa petitioned the International Court of Justice (ICJ) in December 2023, claiming that the Israeli state is committing 'violations of the Genocide Convention'. South Africa's lawyers laid out the grounds on which they are condemning Israel of breaching the 1948 Genocide Convention. UN's Genocide Convention, adopted by the General Assembly on 9 December 1948, was the first human rights treaty to respond to the systematic atrocities committed by the Nazi regime during World War II. The word 'genocide' was first coined by Polish lawyer, Raphael Lemkin, who fled to the USA in 1939 after Germany invaded his country. He combined two words: the Greek prefix *genos*, meaning race or tribe, and the Latin suffix *cide*, meaning killing. The Convention has been ratified by 153 states. The ICJ has repeatedly stated that the Convention embodies principles that are part of general customary international law, meaning that whether or not states have ratified the Genocide Convention, they are all bound as a matter of law by the principle that genocide is a crime prohibited under international law.

Article II of the UN's Genocide Convention explains genocide as

any of the following acts committed with intent to destroy, in whole or in part, a national, ethnical, racial or religious group, as such: (a) Killing members of the group; (b) Causing serious bodily or mental harm to members of the group;(c) Deliberately inflicting on the group conditions of life calculated to bring about its physical destruction in whole or in part; (d) Imposing measures intended to prevent births within the group; (e) Forcibly transferring children of the group to another group.

(UN, 1948)

South Africa's document pointed out to the world that the state of Israel has breached all its responsibilities under international law since 7 October. On 11 and 12 January 2024, the ICJ in the Hague heard South Africa's case claiming that Israel is committing 'genocidal acts' in Gaza. South Africa's 84-page filing to the court makes a case that Israel is generating conditions 'calculated to bring about [Palestinians'] physical destruction'.

Adila Hassim, an advocate of the High Court of South Africa, stood in the court and said that Israel had deliberately imposed conditions that cannot sustain life and that are calculated to bring about the destruction of Gaza through its forced displacement of most of the population.[3] Hassim further said that thousands of families have been displaced multiple times, with half a million now having no homes to return to. She cited how Israel gave entire hospitals orders to evacuate within 24 hours, with no assistance in moving the injured or in moving medical supplies. It did the same with large parts of northern Gaza, where more than 1 million people were asked to move at a short notice. 'The order itself was genocidal', Hassim added (Lawal, 2024). Tembeka Ngcukaitobi, another member of South Africa's legal team, told the ICJ that 'the intent to destroy Gaza has been nurtured at the highest level of state,' and Israel's political and military leaders, including Prime Minister Benjamin Netanyahu, were among 'the genocidal inciters', (Reuters, 2024), and Israel's intent was evident 'from the way in which this military attack is being conducted' (BBC, 2024). South Africa's legal team backed their assertion by referring to some of the incendiary statements by some members of the Israeli government. In November 2023, Israel's heritage minister, Amichai Eliyahu, stated that there was no such thing as non-combatants in Gaza and that dropping a nuclear weapon was an 'option' too (Aldrovandi, 2024).

South Africa's application has been supported by Jordan, Bolivia, Turkey, Malaysia and the Organization of Islamic Cooperation (OIC), which constitutes 57 member states – all are from the Global South (*Global Majority*[4] in a real sense of the term). In January 2024, Indonesia filed a new lawsuit against the Israeli state at the ICJ. In doing so, Indonesia too joined South Africa in taking the Israeli state to the International Court for committing genocide against Palestinians in Gaza (Middle East Monitor, 2024). In addition to this, Mexico and Chile asked the International Criminal Court (ICC) to investigate crimes committed against humanity during the war in Gaza. The ICC was established in 2002 to prosecute individuals for crimes such as genocide, war crimes and other similar offences. ICC deals with individual, not state actions. Israel is not a party to the statute of the court and therefore does not recognise the jurisdiction of the ICC (Financial Times, 2024).

On 17 March 2023, the ICC issued warrants of arrest for Vladimir Putin and Ms. Maria, President of the Russian Federation, and Alekseyevna Lvova-Belova, Commissioner for Children's Rights in the Office of Russian Federation, for the war crime of unlawful deportation of population (children) and that of unlawful transfer of population (children) from occupied areas of Ukraine to the Russian Federation (ICC, 2023). Forced deportation of populations is recognised as a crime under the Rome statute that established the court. Russian Federation was a signatory to the Rome statute but withdrew in 2016, so it has not recognised the jurisdiction of the court since then. Ukraine is not a signatory to the ICC, but it granted the ICC jurisdiction to investigate war crimes committed on its territory.

It seems highly unlikely that President Putin or his Presidential Commissioner for Children's Rights, Lvova-Belova, will surrender to the court's jurisdiction any time soon. However, the issuing of the warrant remains a highly significant moment for the principles of peace and sensitivities about crimes against humanity. The decision of the Court sends a clear signal that no one is immune from prosecution and may serve as a deterrent to military and civilian Russian officials engaged in criminal activity.

The post-World War II period saw many war crimes and serious violations of basic human rights, many of which have gone unpunished, including the USA's use of chemical weapons in Vietnam, the crimes committed by Western colonial powers during the anti-colonial struggles from the 1950s to the 1970s, atrocities during the civil wars in the former Yugoslavia, during the many wars of US-led War on Terror and Russia's wars in Chechnya and Ukraine. This is because enforcing the law is challenging. There is no standing international police force to enforce international law, including the Genocide Convention.

On 26 January 2024, ICJ, the UN's top court, issued an 86-paragraph written decision on the request for provisional measures in the pending case by the government of South Africa accusing Israel of committing genocide in Gaza in violation of the 1948 Genocide Convention. The Court has ordered Israel 'to comply with international law on genocide'. The Court said 'Israel should limit harm to Palestinians in Gaza but stopped short of calling for an immediate end to offensive' (*Financial Times*, 27–28 January 2024: 1). David Kaye in Foreign Affairs stated that the ICJ 'has issued a preliminary ruling in favour of South Africa's claim that Israel's military assault on Gaza may plausibly be characterised as genocide'. Kaye further states that

> the court's ruling also contains a hidden ambition: it challenges all states – and especially the United States – to take international law seriously at a time of increasing violence and conflict and decreasing respect for the authority of international legal institutions.

> The court's order is, despite its apparent moderation, damning. It has allowed litigation to move forward on South Africa's claim that Israel is committing genocide in Gaza, placing a virtual sword of Damocles over not only Israel in its future conduct in Gaza, but also those, such as the United States, that have given it such strong support. It has found plausible South Africa's assertion that Palestinian rights must be protected against genocidal acts. Even Israel's appointee to the court, Judge Aharon Barak, joined the demands that Israel must prevent public and direct incitement to genocide and take 'immediate and effective measures' to enable humanitarian assistance. These are very serious outcomes that reflect global legal; concern about the humanitarian situation in Gaza.

> (Kaye, 2024)

This ruling marks only the beginning of the case. A final ruling by the ICJ on South Africa's genocide claim is likely to take months, even years of litigation over jurisdiction and the ultimate merits of the claim of genocide. Knowing that the ICJ, and ICC, have no real capacity to mandate compliance other than passing their decision to the UN Security Council, where a veto by the USA, and Russia, is almost certain, to have a meaningful impact on the conflict in Gaza is remote. Whatever the court's final determinations might be, the accusation levelled against the Israeli state constitutes a historical watershed with profound symbolic ramifications. A Western-backed democracy has been accused by the ICJ of the most serious international crime that has already made waves. The Israeli state, having accepted the ICJ's legitimacy by arguing the case, will find it very difficult to ignore its ruling. Particularly in the Global South, the case of South Africa's claims is a test of the credibility of the international system. Moreover, a finding of probable genocide would be deeply stigmatising for the collective psyche of the citizens of Israel. The current ICJ proceedings signify an upsetting U-turn of historical record, considering the fact that the genocide crime has now been invoked against Israel, the very state established closely related to the UN Convention and with the same rationale – protecting the Jewish people from a repetition of the genocide. Republic of South Africa's carefully formulated legal procedure against Israel has opened a crack in an increasingly violent international arena and created a light of hope for justice: accountability for perpetrators, redress for victims and hope for peace over violence.

Notes

1 In the Italian original, Gramsci says *fenomeni morbosi*, literally 'morbid phenomena' (Turin: Giulio Einaudi editore, 1977: 311).
2 'The Vice President appears on Meet the Press with Tim Russert', *The White House*, 16 September 2001, https://georgewbush-whitehouse.archives.gov/vicepresident/news-speeches/speeches/vp20010916.html
3 According to a report, collaboratively prepared by researchers from the London School of Hygiene & Tropical Medicine (LSHTM) and the Johns Hopkins Center for Humanitarian Health, on 23 February 2024, claims that 'even in the best-case scenario of an immediate ceasefire there would continue to be thousands of excess deaths after a ceasefire was agreed'. www.lshtm.ac.uk/newsevents/news/2024/report-projects-excess-deaths-due-gaza-crisis
4 *Global Majority* is a collective term that first and foremost speaks to and encourages those so-called to think of themselves as belonging to the majority on planet Earth. It refers to people who are Black, African, Asian, Brown, dual-heritage, indigenous to the Global South and/or have been racialised as 'ethnic minorities'. Globally, these groups currently represent more than 80 per cent of the world's population, making them the global majority now, and with current growth rates, they are set to remain so for the foreseeable future (Campbell-Stephens, 2020).

References

ABC News (2003) 'Were 1998 memos a blueprint for war?', 8 March. Available at: https://abcnews.go.com/Nightline/story?id=128491&page=1

Abdul-Ahad, G. (2019) 'Iraq's young protesters count cost of a month of violence', *The Guardian*, 29 October. Available at: www.theguardian.com/world/2019/oct/29/iraqi-protesters-demonstrations-month-of-violence

Abelow, B. (2022) *How the West brought war to Ukraine*. Massachusetts: Siland Press.

Action on Armed Violence (2023) 'Ukraine: AOAV explosive violence data on harm to civilians'. Available at: https://aoav.org.uk/2023/ukraine-casualty-monitor/

Action on Armed Violence (2024a) '122% rise in global civilian fatalities from explosive weapons in 2023: A year of harm reviewed'. Available at: https://aoav.org.uk/2024/2023-a-year-of-explosive-violence-harm-reviewed/.

Action on Armed Violence (2024b) 'Briefing note: Explosive weapons use in Yemen, 2014–2023'. Available at: https://aoav.org.uk/2024/briefing-note-explosive-weapons-use-in-yemen-2014-2023/

Adams, Paul and Hancock, Sam (2023) 'Ben Wallace: Ukraine has 'tragically become a battle lab' for war technology', *BBC*, July 18 [online] Available at: www.bbc.co.uk/news/uk-66229336

Ahmad, M.I. (2015) 'Death from above, remotely controlled: Obama's drone wars', *In These Times*. Available at: https://inthesetimes.com/article/drones-andrew-cockburn-kill-chain-chris-woods-sudden-justice

Airwars (2024) 'Russian military in Syria'. Available at: https://airwars.org/conflict/russian-military-in-syria/

Al Jazeera (2019) 'Is there hope for peace in Afghanistan?' Available at: www.aljazeera.com/programmes/insidestory/2019/01/hope-peaceafghanistan-190118171234537.htm(Accessed 2 June 2024).

Al Jazeera (2022). 'Ukraine-Russia crisis: What is the Minsk agreement?', 9 February. Available at: www.aljazeera.com/news/2022/2/9/what-is-the-minsk-agreement-and-why-is-it-relevant-now

Aldrovandi, Karlo (2024). 'Gaza war: How South Africa's genocide case against Israel is shaping up', Trinity College-Dublin, 17 January. Available at: www.tcd.ie/news_events/top-stories/featured/the-conversation-gaza-war-how-south-africas-genocide-case-against-israel-is-shaping-up/ (Accessed 2 June 2024).

Aleem, Zeeshan (2022). 'Russia's Ukraine invasion may have been preventable', *MSNBC*, 4 March. Available at: www.msnbc.com/opinion/msnbc-opinion/russia-s-ukraine-invasion-may-have-been-preventable-n1290831

Alfonso, Fernando (2021) *CNN* . Available at: https://edition.cnn.com/world/live-news/afgh anistan-kabul-taliban-us-news-08-28-21/h_675be4324156edc80ef667536defe716

Alkarama (2022) 'Iraq Shadow report', 20 January. Available at: www.alkarama.org/sites/default/files/2022-02/INT_CCPR_CSS_IRQ_47787_E.pdf

Allan, Duncan (2020). 'The Minsk Conundrum: Western policy and Russia's war in Eastern Ukraine', Chatham House, 22 May. Available at: www.chathamhouse.org/2020/05/minsk-conundrum-western-policy-and-russias-war-eastern-ukraine-0/minsk-implementation

Allison, Graham (2015). 'The Thucydides Trap: Are the U.S. and China headed for war?', *The Atlantic*, 24 September. Available at: www.theatlantic.com/international/archive/2015/09/united-states-china-war-thucydides-trap/406756/

Anderson, B. (1983) *Imagined communities: Reflections of the origin and spread of nationalism*. London: Verso.

Anderson, Kenneth (2010) 'Rise of the drones: Unmanned systems and the future of war', *Written Testimony Submitted to Subcommittee on National Security and Foreign Affairs, Committee on Oversight and Government Reform, US House of Representatives, Subcommittee Hearing,* March 23.

Antonov, Anatoly (2021) 'An existential threat to Europe's security architecture?', *Foreign Policy* [online] Available at: https://foreignpolicy.com/2021/12/30/russia-ukraine-nato-threat-security/

Arhirova, Hanna (2023) 'Russia launches more drone attacks as Zelensky travels to a European forum', *ABC News*. Available at: https://abcnews.go.com/International/wireSt ory/russia-launches-drone-attacks-ukraine-president-zelenksyy-travels-103745415 (Accessed 15 January 2024).

Arrighi, Giovanni and Beverly J. Silver (2001). 'Capitalism and world (dis)order', *Review of International Studies*, 27, special issue: *Empires, Systems and States: Great Transformations* in International Politics, December, pp. 257–279. Available at: https://library.fes.de/libalt/journals/swetsfulltext/16957009.pdf

Ash, T.G. (2001) 'A day to define a century', *openDemocracy*, 12 September. Available at: www.opendemocracy.net/en/article_122jsp/

Babayan, N. (2017) 'Bearing truthiness: Russia's cyclical legitimation of its actions', *Europe-Asia Studies*, 69(7): 1090–1105.

Bacon, Edwin (2018). 'Policy change and the narratives of Russia's think tanks', *Palgrave Communications*, 4(94), www.nature.com/articles/s41599-018-0148-y

Baker, P. (2023) 'In unforgiving terms, Biden condemns "evil" and "abhorrent" attack on Israel', *The New York Times*, 10 October. Available at: www.nytimes.com/2023/10/10/us/politics/biden-israel-hamas.html

Bandow, D. (2021) 'War is big business'. *CATO Institute*. Available at: https://www.cato.org/commentary/war-big-business

Banerjee, N and S. Tavernise (2001) 'As the war shifts alliances, oil deals follow', *The New York Times*, 15 December. Available at: www.nytimes.com/2001/12/15/business/as-the-war-shifts-alliances-oil-deals-follow.html

BBC (2016) 'UK military deaths in Iraq'. Available at: www.bbc.co.uk/news/uk-10637526

BBC (2017) 'Afghanistan war: Trump's allies and troop numbers'. Available at: www.bbc.co.uk/news/world-41014263

BBC (2020) 'Afghan maternity ward attack'. Available at: www.bbc.co.uk/news/world-asia-52675705

BBC (2022). 'IMF: Ukraine economy could shrink as much as 35%', *BBC News*, 14 March. Available at: www.bbc.co.uk/news/business-60743592

Bellamy, Alex J. (2023). *Warmonger: Vladimir Putin's imperial wars*. Newcastle upon Tyne: Agenda.

Bergen, Sterman and Salyk-Virk (2021) 'America's Counterterrorism Wars'. *New America*, 17 June. Available at: www.newamerica.org/future-security/reports/americas-counterte rrorism-wars/

Bignell, P., A. McSmith and J. Brown (2011) 'Iraqi oil supply was considered to be "vital" to British interests', *Independent*, 20 April. Available at: www.independent.co.uk/news/ uk/politics/iraqi-oil-supply-was-considered-to-be-vital-to-british-interests-2270072.html

Bogner, M. (2022*). Ukraine: Monitoring the devastating impact of the war on civilians*. Kyiv: United Nations.

Bondar, K. (2023) 'Arsenal of democracy: Integrating Ukraine into the West's defense industrial base', Carnegie Endowment for International Peace. Available at: https://carneg ieendowment.org/2023/12/04/arsenal-of-democracy-integrating-ukraine-into-west-s-defense-industrial-base-pub-91150 (Accessed 5 December 2023)

Bourgois, Pierre (2020) 'The PNAC (1997–2006) and the post-Cold War "neoconservative moment"', *E-International Relations*, 1 February. Available at: www.e-ir.info/2020/02/ 01/new-american-century-1997-2006-and-the-post-cold-war-neoconservative-moment/

Brisard, Jean-Charles and Dasquie, Guillaume (2002) *Forbidden truth: US-Taliban secret oil diplomacy and the failed hunt for Bin Laden*, trans. Lucy Rounds with Peter Fifield and Nicholas Greenslade. New York: Thunder's mouth Press/Nation Books.

Brown, Dee (1970) *Bury my heart at wounded knee*. New York: Henry Holt & Company.

Brussels Summit Communiqué (2021) Press Release June 14, NATO [online] Available at: www.nato.int/cps/en/natohq/news_185000.htm.

Brzezinski Interview with *Le Nouvel Observateur* (1998) in David N. Gibbs, https://dgibbs. arizona.edu/content/brzezinski-interview-2

Brzezinski, Z. (1997) *The Grand Chessboard: American Primacy and its Geostrategic Imperatives*. New York: Basic Books.

Buchanan, I. and L. Guillaume (2009) 'The spectacle of war: Security, legitimacy and profit post-9/11', in R. Braidotto, C. Colebrook and P. Hanafin, eds. *Deleuze and law: Forensic futures*. Palgrave Macmillan: 179–197.

Buchholz, Katharina (2022) 'The world's biggest arms exporters' [online] Available at: www.statista.com/chart/18417/global-weapons-exports/

Bush, Laura (2001) 'Radio address by Mrs. Bush', *Office of the First Lady*, 17 November. Available at: https://georgewbush-whitehouse.archives.gov/news/releases/2001/11/20011 117.html

Calabresi, Massimo (2014). 'Inside Putin's East European Spy Campaign', *Time*, 7 May. Available at: https://time.com/90752/inside-putins-east-european-spy-campaign/

Campbell-Stephens, R. (2020) 'Global Majority; Decolonising the language and Reframing the Conversation about Race'. Available at: https://www.leedsbeckett.ac.uk/-/media/files/ schools/school-of-education/final-leeds-beckett-1102-global-majority.pdf

Carr, E.H. (1981) *The twenty years crisis, 1919–1939*. London: Macmillan.

Chilcot Report (2016) 'The report of the Iraq Inquiry', 6 July. Available at: https://assets. publishing.service.gov.uk/media/5a80f42ced915d74e6231626/The_Report_of_the_Ir aq_Inquiry_-_Executive_Summary.pdf

Children of War (2023) Available at: https://childrenofwar.gov.ua/en/

Children of War (2023) Children's stories. Available at: https://childrenofwar.gov.ua/en/stor ies-of-children/

Cirincione, Joseph (2003) 'Origins of regime change in Iraq', *Carnegie*, 19 March. Available at: https://carnegieendowment.org/2003/03/19/origins-of-regime-change-in-iraq-pub-1214 (Accessed 2 June 2024).

Clausewitz, Carl Von (1982) *On war.* London: Penguin.

Clover, C. (1999) 'Dreams of the Eurasian Heartland: The Reemergence of Geopolitics', *Foreign Affairs*, 78, March–April. Available at: www.foreignaffairs.com/articles/asia/1999-03-01/dreams-eurasian-heartland-reemergence-geopolitics

CNN (2011) https://edition.cnn.com/2011/10/21/world/meast/chart-us-troops-iraq/index.html

Coates, A.J. (2016) *The ethics of war.* Manchester: Manchester University Press.

Cockburn, A. (2015) *Kill chain: Drones and the rise of high-tech assassins.* London, New York: Verso.

Cockburn, A. and James Ridgeway (1981) 'The world of appearance: The public campaign', in Thomas Ferguson and Joel Rogers, eds. *The hidden election: Politics and economics in the 1980 presidential campaign.* New York: Pantheon Books.

Cohen, Stephen F. (2010). 'An interview with Stephen Cohen', *Journal of International Affairs*, Spring/Summer, 63(2): 191–205, https://ciaotest.cc.columbia.edu/journals/jia/v63i2/f_0022003_18160.pdf

Cohen, Stephen F. (2017). 'The New Cold War is already more dangerous was its predecessor', *The Nation*, 11 October. Available at: www.thenation.com/article/archive/the-new-cold-war-is-already-more-dangerous-than-was-its-predecessor/

Congress (2001) 'Public Law 107–140: 107th Congress Joint Resolution'. Available at: www.congress.gov/107/plaws/publ40/PLAW-107publ40.pdf

Congressional Research Service (2023) 'U.S. security assistance to Ukraine' [online] Available at: https://crsreports.congress.gov/product/pdf/IF/IF12040?loclr=bloploc

Connah, Leonie (2020) 'US intervention in Afghanistan: Justifying the unjustifiable?', *South Asia Research*, 41(1): 70–86. Available at: https://journals.sagepub.com/doi/pdf/10.1177/0262728020964609

Cook, Samuel J. (2015) 'The Crimean crisis and international law', *University of St Thomas Journal of Law and Public Policy*, 10(1), Fall. Available at: https://core.ac.uk/reader/217155680

Correlates of War 'COW War Data, 1816–2007' Available at: https://correlatesofwar.org/data-sets/cow-war/

Cortright, D. (2011) *Ending Obama's war: Responsible military withdrawal from Afghanistan.* Boulder: Paradigm.

Courtney, W. (2023) 'Russia's appetite may extend beyond Ukraine'. Available at: www.rand.org/pubs/commentary/2023/02/russias-appetite-may-extend-beyond-ukraine.html (Accessed 2 June 2024)

Cranny-Evans, Sam (2022). 'The Chechens: Putin's loyal foot soldiers', *RUSI*, 4 November, Available at: www.rusi.org/explore-our-research/publications/commentary/chechens-putins-loyal-foot-soldiers

Crisis Group (2024). 'Conflict in Ukraine's Donbas: A visual explainer'. Available at: www.crisisgroup.org/content/conflict-ukraines-donbas-visual-explainer

Dalio, Ray (2023) 'A US-Chinese led world peace?', 16 October. Available at: www.linkedin.com/pulse/us-chinese-led-world-peace-ray-dalio/

D'Anieri, P. (2019) *Ukraine and Russia: From civilized divorce to uncivil war.* Cambridge: Cambridge University Press.

Dardagan, H., Hamourtziadou, L. and Sloboda, J. (2023) 'Iraq's residual war, Iraq Body Count'. Available at: www.iraqbodycount.org/analysis/beyond/residual-war/

De Roose, F. (1990) 'Self-defence and national defence', *Journal of Applied Philosophy*, 7(2): 159–168.

Del Ponte, C. and Blewitt, G. (2023) 'International justice must serve victims of Israel-Hamas war atrocities', *Politico*. Available at: www.politico.eu/article/israel-hamas-atrocit ies-victims-justice-icc-hague/

Dicken, P. (2007) *Global shift: Mapping the changing contours of the world economy*. First edition. London: Sage.

Doak, E. (2023) 'The Iraq war at twenty', *The American Conservative*, 17 March. Available at: www.theamericanconservative.com/the-iraq-war-at-twenty/

Dobbs, M. and Goshko, J.M. (1996) 'Albright's personal odyssey shaped foreign policy beliefs', *Washington Post*, 6 December. Available at: www.washingtonpost.com/archive/ politics/1996/12/06/albrights-personal-odyssey-shaped-foreign-policy-beliefs/fdef6884-c540-451f-8b91-f07b211645cc/

Eichergreen, Barry (2001) 'US foreign economic policy after 11 September', *Social Science Research Council*, After 11 September. Available at: www.ssrc.org/sept11/essays/eich ergreen.

End Cross-border Bombing (2022) 'Civilian casualties of Turkish military operations in Northern Iraq (2015–2021)'. Available at: https://cpt.org/wp-content/uploads/ECBB-Rep ort-en.pdf

Euro-Med Human Rights Monitor (2024) 'Statistics on the Israeli attack on the Gaza Strip (7 October–3 February 2024)'. Available at: https://euromedmonitor.org/en/article/6135/Sta tistics-on-the-Israeli-attack-on-the-Gaza-Strip-%2807-October---03-February-2024%29

European Central Bank (2005) 'Review of the international role of the Euro', European Central Bank, January. Available at: www.ecb.europa.eu/pub/pdf/other/euro-internatio nal-role2005en.pdf

European Commission (2023) 'Ukraine: International centre for the prosecution of Russia's crime of aggression against Ukraine starts operations today'. Available at: https://neighb ourhood-enlargement.ec.europa.eu/news/ukraine-international-centre-prosecution-russ ias-crime-aggression-against-ukraine-starts-operations-2023-07-03_en

Every Casualty Counts, 'Our mission'. Available at: https://everycasualty.org/

Fadel, L (2010) 'U.S. reports 77,000 Iraqi fatalities from 2004 to August 2008', *Washington Post*, 15 October. Available at: www.washingtonpost.com/wp-dyn/content/article/2010/ 10/14/AR2010101406139.html

Fathi, R.A., Matti, L.Y., Al-Salih, H.S. and Godbold, D. (2013) 'Environmental pollution by depleted uranium in Iraq with special reference to Mosul and possible effects on cancer and birth defect rates', *Medicine, Conflict and Survival*, 29(1): 7–25.

Faure, J. (2022) 'The deep ideological roots of Russia's war', *Le Monde Diplomatique*, April. Available at: https://mondediplo.com/2022/04/03ideology

Faure, J. (2023) 'What role did ideology play in triggering Russia's invasion of Ukraine', *The Russia Programme*, October. Available at: https://therussiaprogram.org/onlinepaper_8

Ferguson, T. and J. Roger (1987) *Right turn: The decline of the democrats and the future of American politics*. Chicago: Lawrence Hill Books.

Financial Times (2024). 'ICC asked to probe Israel-Hamas war', 20 January. Available at: www.ft.com/content/77bddb7d-87b2-4c4e-a856-10c2e13f253a

Fordham, A. (2021) 'In Iraq's "dire" economy, poverty is rising – and so are fears of instability', *NPR*. Available at: www.npr.org/2021/02/03/961149079/in-iraqs-dire-econ omy-poverty-is-rising-and-so-are-fears-of-instability?t=1627992007423

Fouskas, V. and B Gokay (2005) *The new American imperialism. Bush's war on terror and blood for oil*. Westport: Praeger Security International.

Fouskas, V. and B Gokay (2012) *The fall of the US empire. Global fault-lines and the shifting imperial order*. London: Pluto Press.

Frank, Andre Gunder (1999) 'NATO, Caucasus/Central Asia oil', *WSWS*. Available at: www.wsws.org/index.html (Accessed 16 June).

Freeman, C. (2022) 'Pushback with Aaron Maté' [podcast and video] March 24 [online] Available at: https://thegrayzone.com/2022/03/24/us-fighting-russia-to-the-last-ukrainian-veteran-us-diplomat/

Frew, Joanna (2018) *Drone wars: The next generation*. Oxford: Drone Wars UK. Available at: https://dronewars.net/wp-content/uploads/2018/05/dw-nextgeneration-web.pdf

Fukuyama, Francis (1989) 'The end of history?', *The National Interest*, No. 16. pp 3–18, Summer. Available at: https://www.jstor.org/stable/24027184

Fukuyama, Francis (1992) *The end of history and the last man*. New York: The Free Press.

Garamone (2017) Available at: www.defense.gov/News/News-Stories/Article/Article/1255158/special-ops-capabilities-relevant-around-the-world-commander-says/

Gaza Martyrs twitter account (2023) Available at: https://twitter.com/GazaMartyrs/status/1734995650452885594

Gilmore, Gerry J. (2003) 'Franks: Iraq campaign is "unlike any other in history"', *American Forces Press Service*, 22 March. Available at: www.af.mil/News/Article-Display/Article/139761/franks-iraq-campaign-is-unlike-any-other-in-history/

Gilpin, R. (1981) *War and change in world politics*. Cambridge: Cambridge University Press.

Global Terrorism Database (2022) Available at: www.start.umd.edu/gtd/

Global Terrorism Database incident (5 August 2015) 201508050066 Available at: www.start.umd.edu/gtd/search/IncidentSummary.aspx?gtdid=201508050066

Global Terrorism Database incident (3 August 2015) 201508020073 Available at: www.start.umd.edu/gtd/search/IncidentSummary.aspx?gtdid=201508020073

Gokay, Bulent (2016) 'Why you can't explain the Iraq War without mentioning oil', *Conversation*, 8 July. Available at: https://theconversation.com/why-you-cant-explain-the-iraq-war-without-mentioning-oil-59352

Gokay, Bulent (2022) 'America's longest war was founded on false pretences', *Journal of Global Faultlines*, 9(1). pp. 5–8

Gokay, B and R.B. J. Walker eds. (2003) *09/11/2001: War, terror and judgement*. London: Frank Cass.

Gonzalez, Roberto J. (2023) 'Drones over Ukraine: What the war means for the future of remotely piloted aircraft in combat', The Conversation [online] Available at: https://theconversation.com/drones-over-ukraine-what-the-war-means-for-the-future-of-remotely-piloted-aircraft-in-combat-197612

Gordon, Michael R., Pancevski, Bojan, Bisserbe, Noemie and Walker, Marcus (2022) 'Vladimir Putin's 20-year march to Ukraine – and how the West mishandled it', *The Wall Street Journal*, 1 April [online] Available at: www.wsj.com/articles/vladimir-putins-20-year-march-to-war-in-ukraineand-how-the-west-mishandled-it-11648826461 (Accessed 2 June 2024).

Gossman, P. (2018) 'Another airstrike in Kunduz, and more civilian deaths'. Available at: www.hrw.org/news/2018/04/11/another-airstrike-kunduzand-more-civilian-deaths. (Accessed 2 June 2024).

Government.ru (2013) 'Federal'naya tselevaya programma: ukreplenie edinstva rossiiskoi natsii I etnokul'turnoe razvitie narodov Rossii (2014–2020 gg)', Government.ru, 20 August. Available at: http://government.ru/media/files/41d4862001ad2a4e5359.pdf

Gowan, P. (2004). Contemporary intra-core relations and world systems theory. *Journal of World-Systems Research, 10* (2), 471–500. Available at: https://jwsr.pitt.edu/ojs/jwsr/article/view/291

Gowan, P. (2002). A calculus of power. *New Left Review, 16.* Available at http://newleftreview.org/II/16/peter-gowan-a-calculus-of-power.

Gramsci, Antonio (1971). *Selections from the prison notebooks of Antonio Gramsci*, ed. and translated by Quintin Hoare and Geoffrey Nowell-Smith. London: Lawrence & Wishart.

Grassley, Chuck (2023) 'Putin is an imperialist who must be stopped now or he will become more dangerous'. Available at: www.grassley.senate.gov/news/remarks/putin-is-an-imperialist-who-must-be-stopped-now-or-he-will-become-more-dangerous

Green, C. (2017) *Spin Zhira: Old man in Helmand*, 3rd edition. London: OMiH.

Griffiths, James (2015) 'Collateral damage: A brief history of U.S. mistakes at war', 7 October. Available at: https://edition.cnn.com/2015/10/06/middleeast/us-collateral-damage-history/index.html

Haaretz (2023) 'Israel's dead: The names of those killed in Hamas attacks, massacres and the Israel-Hamas war'. Available at: www.haaretz.com/israel-news/2023-10-19/ty-article-magazine/israels-dead-the-names-of-those-killed-in-hamas-massacres-and-the-israel-hamas-war/0000018b-325c-d450-a3af-7b5cf0210000

Hallin, Daniel C. (1989) *The uncensored war: The media and Vietnam.* Berkeley: University of California Press.

Hamourtziadou, L. (2014) 'Fault lines to trenches: Iraq 2003–2014', *Journal of Global Faultlines*, 2(1): 98–108. Available at: www.scienceopen.com/hosted-document?doi=10.13169/jglobfaul.2.1.0098

Hamourtziadou, L (2021) *Body count: The War on Terror and civilian deaths in Iraq.* Bristol: Bristol University Press.

Hamourtziadou, L. (2023) 'Independent inquiry relating to Afghanistan: day 3', Action on Armed Violence. Available at: https://aoav.org.uk/2023/independent-inquiry-relating-to-afghanistan-day-3/

Hamourtziadou, L. (2024) *The ethics of remote warfare.* Cardiff: University of Wales Press.

Hamourtziadou, L., Dardagan, H. and Sloboda, J. (2019) 'Iraq in 2019: Calls for a 'True Homeland' met with deadly violence, Iraq Body Count'. Available at: www.iraqbodycount.org/analysis/numbers/2019/

Hamourtziadou, L. and Gokay, B. (2020) 'Iraq's security 2003–2019: Death and neoliberal destruction par excellence', *openDemocracy.* Available at: www.opendemocracy.net/en/north-africa-west-asia/iraqs-security-2003-2019-death-and-neoliberal-destruction-par-excellence/

Hamourtziadou, L. and Gokay, B. (2021) 'The deadly legacy of 20 years of US "War on Terror' in Iraq", *openDemocracy.* Available at: www.opendemocracy.net/en/north-africa-west-asia/the-impact-of-the-War on Terror-on-iraq-state-economy-and-civilian-deaths/.

Hamourtziadou, L. and Jackson, J. (2020) 'Covid 19 and the myth of security', *Journal of Global Faultlines*, 7(1): 96–98. Available at: www.scienceopen.com/hosted-document?doi=10.13169/jglobfaul.7.1.0096

Harding, L. (2022) *Invasion. Russia's bloody war and Ukraine's fight for survival.* London: Guardian Faber.

Harmash, Olena (2023) 'Russia's biggest drone strike in weeks hits Ukrainian infrastructure', Reuters. Available at: www.reuters.com/world/europe/russian-drones-hit-civilian-target-ukraines-kharkiv-officials-say-2023-11-02/ (Accessed 2 June 2024).

Harvey, D. (2005) *A brief history of neoliberalism.* Oxford: Oxford University Press.

Haslam, Jonathan (2022) 'Russia attacks Ukraine: A post mortem for Putin', *H-Diplo*, 17 March. Available at: https://networks.h-net.org/node/28443/discussions/9948394/h-diplo-essay-420-commentary-series-putin's-war-"russia-attacks

Haslam, J. (2011) *Russia's Cold War: From the October Revolution to the Fall of the Wall.* New Haven: Yale University Press.

Held, D. (2001) 'Violence and justice in a global age', *openDemocracy*, 14 September. Available at: www.opendemocracy.net/en/article_144jsp/

Herman, Steve (2022) Voa News, 8 March. Available at: www.voanews.com/a/is-putin-the-new-hitler-/6476408.html

Hideg, G. (2023) 'From conflict to consequence: Nearly half of Ukrainian men would like to own a firearm, or already have one', Small Arms Survey. Available at: www.smallarmssurvey.org/resource/conflict-consequence-ukraine.

Hobsbawm, Eric J. (1994) *The age of extremes: A history of the world, 1914–1991.* New York: Vintage.

Holmes, R.L. (1989) *On war and morality.* Princeton: Princeton University Press.

Hoover Institution (1999) 1 February. www.hoover.org/research/using-power-and-diplomacy-deal-rogue-states

Hulsemann, Laura (2023) 'Putin could attack Baltics and Moldova next, says Belgian army chief', *Politico*, 19 December. Available at: www.politico.eu/article/belgian-army-chief-hofman-putin-attack-after-ukraine-baltics-moldova-next-russia/

Human Rights Watch (2022) 'Questions and answers: Turkey's threatened incursion into Northern Syria'. Available at: www.hrw.org/news/2022/08/17/questions-and-answers-turkeys-threatened-incursion-northern-syria#Q3

Human Security Now (2003) 'Human Security Now: Protecting and empowering people'. Available at: https://reliefweb.int/report/world/human-security-now-protecting-and-empowering-people

Huntington, Samuel P. (1997) *The clash of civilizations and the remaking of world order.* New York: Simon and Shuster.

Independent (2022) 'Putin is "Hitler of the 21st century", says Ireland's Leo Varadkar', 25 February. Available at: www.independent.co.uk/news/world/europe/putin-russia-ukraine-varadkar-hitler-b2023656.html

Institute for Economics and Peace (2018) 'Global Terrorism Index 2018: Measuring the impact of terrorism'. Available at: https://reliefweb.int/report/world/global-terrorism-index-2018#:~:text=This%20year's%20Global%20Terrorism%20Index,occurring%20in%20Iraq%20and%20Syria.

International Criminal Court (2020) 'Situation in Iraq/UK final report', The Office of the Prosecutor. Available at: www.icc-cpi.int/sites/default/files/itemsDocuments/201209-otp-final-report-iraq-uk-eng.pdf

International Criminal Court (2021) 'Statement by the ICC Prosecutor, Ms. Fatou Bensouda, on an investigation into the situation in Palestine', 3 March. Available at: www.icc-cpi.int/fr/news/declaration-du-procureur-de-la-cpi-mme-fatou-bensouda-propos-dune-enquete-sur-la-situation-en

International Criminal Court (ICC) (2023). 'ICC press release: Situation in Ukraine', 17 March. Available at: www.icc-cpi.int/news/situation-ukraine-icc-judges-issue-arrest-warrants-against-vladimir-vladimirovich-putin-and

International Criminal Court, Cases (2024). Available at: www.icc-cpi.int/cases

International Physicians for the Prevention of Nuclear War, Physicians for Social Responsibility, and Physicians for Global Survival, eds (2015) 'Body count: Casualty figures after 10 years of the "War on Terror"'. International edition – Washington DC, Berlin, Ottawa – March, translated from German by Ali Fathollah-Nejad.

Iraq Body Count (2005) 'A dossier of civilian casualties 2003–2005'. Available at: www.iraqbodycount.org/analysis/reference/pdf/a_dossier_of_civilian_casualties_2 003-2005.pdf

Iraq Body Count (2010) 'Iraq war logs: The truth is in the details'. Available at: www.iraqbo dycount.org/analysis/beyond/warlogs/

Iraq Body Count (2011) 'The unexamined Iraqi dimension of UK involvement in Iraq'. Available at: www.iraqbodycount.org/analysis/beyond/uk-involvement/

Iraq Body Count, Twitter (2023) 'First weeks of civilian deaths in Iraq (2003) and Gaza (2023) compared by day number'. Available at: https://twitter.com/iraqbodycount/status/ 1719417486133678413

Iraq Body Count (2023) 'Names of persons killed in Gaza extracted from Health Ministry list released 2023-10-26, machine transliterated from Arabic into English'. Available at: www.iraqbodycount.org/pal/gaza-names-of-killed.xlsx. Original data from: www.pal estinechronicle.com/wp-content/uploads/2023/10/here.pdf

Iraq Body Count incident j036-i Available at: www.iraqbodycount.org/database/incidents/ j036-i

Iraq Body Count incident a6262 Available at: www.iraqbodycount.org/database/incide nts/a6262

Iraq Body Count incident x025 Available at: www.iraqbodycount.org/database/incide nts/x025

Iraq Body Count incident j038 Available at: www.iraqbodycount.org/database/incide nts/j038

Iraq Body Count incident d3412 Available at: www.iraqbodycount.org/database/incide nts/d3412

Iraq Body Count incident k4435 Available at: www.iraqbodycount.org/database/incide nts/k4435

Iraq Body Count incident a6079 Available at: www.iraqbodycount.org/database/incide nts/a6079

Iraq Body Count incident k1099 Available at: www.iraqbodycount.org/database/incide nts/k1099

Iraq Body Count incident d12838 Available at: www.iraqbodycount.org/database/incidents/ d12838

Iraq Body Count incident d12801 Available at: www.iraqbodycount.org/database/incidents/ d12801

Iraq Body Count incident a5497 Available at: www.iraqbodycount.org/database/incide nts/a5497

Iraq Body Count incident k001 Available at: www.iraqbodycount.org/database/incide nts/k001

Iraq Body Count incident a6250 Available at: www.iraqbodycount.org/database/incide nts/a6250

Iraq Body Count incident a5885 Available at: www.iraqbodycount.org/database/incide nts/a5885

Iraq Body Count Monthly Civilian Deaths (2003–2013) Available at: www.iraqbodycount. org/database/

Iraq Body Count: The ISIS Years (monthly civilian deaths 2014–2017) Available at: www. iraqbodycount.org/database/

Juhasz, A. (2013) 'Why the war in Iraq was fought for Big Oil', CNN Opinion, 15 April. Available at: https://edition.cnn.com/2013/03/19/opinion/iraq-war-oil-juhasz/index.html

Kagan, Robert (2004) *Paradise and power: America and Europe in the New World Order*. London: Atlantic Books.

Kaldor, M. (2007) *Human security*. Cambridge: Polity Press.

Kaldor, M. (2013) *New and old wars: Organized violence in a global era*, 3rd edition. Hoboken: Wiley Publications.

Kaye, David (2024). 'The ICJ ruling's hidden democracy', *Foreign Affairs*, 26 January. Available at: www.foreignaffairs.com/israel/icj-rulings-hidden-diplomacy

Kennedy, P. (1987) *The rise and fall of great powers: Economic change and military conflict from 1500 to 2000*. New York: Random House.

Kenworthy, E.W. (1961) 'U.S. to Help Saigon Fight Reds With More Experts and Planes; SAIGON WILL GET MORE U.S. HELP'

The New York Times, November 17. Available at: www.nytimes.com/1961/11/17/archives/us-to-help-saigon-fight-reds-with-more-experts-and-planes-saigon.html?auth=login-goo gle1tap&login=google1tap

Kerbel, Matthew Robert (1998) 'Parties in the media: Elephants, donkeys, boars, pigs, and jackals', in L. Sandy Maisal, *The parties respond: Changes in American parties and campaigns*, 3rd edition. Boulder: Westview Press.

Khylko, Maksym and Oleksandr Tytarchuk (2017) *Human security and security sector reform in eastern Europe*. Kyiv: EESRI. Available at: https://library.fes.de/pdf-files/bue ros/ukraine/13420.pdf

King's Centre for Military Health Research and Academic Department of Military Mental Health (2023) 'The mental health and wellbeing of the UK armed forces community'. Available at: https://kcmhr.org/key-facts/

Kissinger, Henry (2001) 'Foreign policy in the age of terrorism', transcript of the 2001 *Ruttenberg Lecture*, The Centre for Policy Studies.

Kolesnikova, N (2023) 'Civilians killed in Russia, Ukraine; more drones hit Moscow', Vanguard. Available at: www.vanguardngr.com/2023/08/civilians-killed-in-russia-ukra ine-moscow-hit-by-more-drones/

KSE-Kyiv School of Economics (2023) 'The total amount of damage caused to the infrastructure of Ukraine due to the war reaches $151.2 billion – estimate as of September 1, 2023', 3 October. Available at: https://kse.ua/about-the-school/news/the-total-amount-of-damage-caused-to-the-infrastructure-of-ukraine-due-to-the-war-reaches-151-2-billion-estimate-as-of-september-1-2023/

Kunertova, D. (2023) 'The war in Ukraine shows the game-changing effect of drones depends on the game', *Bulletin of the Atomic Scientists*, 79(2): 95–102 [online] Available at: www.tandfonline.com/doi/full/10.1080/00963402.2023.2178180

Lafta, R.K. and Merza, A.K. (2021) 'Women's mental health in Iraq post-conflict', *Med Confl Surviv*, 37(2): 146–159. Available at: https://pubmed.ncbi.nlm.nih.gov/34182837/

Landes, D. (1969) *The unbound Prometheus: Technological change and industrial development in Western Europe from 1750 to the present*. Cambridge: Cambridge University Press.

Lang, Anthony F.J.R., O'Driscoll, Cian and Williams, John, eds. (2013) *Just war: Authority, tradition, and practice* . Washington: Georgetown University Press.

Laruelle, Marlene (2016) 'The Izborsky club, or the new conservative avant-garde in Russia', *The Russian Review*, 75(4): 626–644.

Laruelle, Marlene (2022) 'What is the ideology of a mobilized Russia', *Russia.Post*, 4 October. Available at: https://russiapost.info/politics/war_ideology

Lawall, Shola (2024), 'What is South Africa's five-point ICJ argument against Israel', *Al Jazeera*, 12 January. Available at: www.aljazeera.com/news/2024/1/12/icj-genocide-case-south-africas-five-point-argument-against-israel

Leipold, J.D. (2015) 'Milley: Russia no.1 threat to US', *US Army*, 10 November. Available at: www.army.mil/article/158386/milley_russia_no_1_threat_to_us

Leoni, Zeno (2023) *Grand strategy and the rise of China*. Newcastle upon Tyne: Agenda.

Lieven, Anatol (2001) 'The end of NATO', *Prospect*, December. Available at: www.prospectmagazine.co.uk/opinions/56099/the-end-of-nato

Ling, J. (2022) 'How to prosecute Russia's war crimes', *Foreign Policy*, 12 August. Available at: https://foreignpolicy.com/2022/08/12/how-to-prosecute-russias-war-crimes/

Loescher, G. (2002). 'Blaming the victim: Refugees and global security'. *Bulletin of the Atomic Scientists*, 58(6): 46–53.

Looney, Robert (2003) 'The neoliberal model's planned role in Iraq's economic transition', *Middle East Journal*, 57(4), Autumn. Available at: www.jstor.org/stable/4329940

Loveluck, L. (2021) 'Iraqi authorities denying prisoners their rights from arrest to prosecution, U.N. says', *The Washington Post*, 3 August. Available at: www.washingtonpost.com/world/iraq-authorities-deny-prisoner-rights/2021/08/03/5f9d2c7e-f07a-11eb-81b2-9b7061a582d8_story.html

Loveluck, L., George, S. and Birmbaum, M. (2023) 'As Gaza's death toll soars, secrecy shrouds Israel's targeting process', *The Washington Post*, 5 November. Available at: www.washingtonpost.com/world/2023/11/05/israel-strike-targets-gaza-civilians-hamas/

Lubold, Gordon and Nancy Youssef (2017) 'U.S. has more troops in Afghanistan than publicly disclosed', *Wall Street Journal*, 22 August. Available at: www.wsj.com/articles/u-s-has-more-troops-in-afghanistan-than-publicly-disclosed-1503444713

Lynch, Lily (2023) 'The realists were right', *the New Statesman*, 2 September. Available at: www.newstatesman.com/the-weekend-essay/2023/09/ukraine-war-realists-right

MacDonald, A. and Sivorka, I. (2023) 'The race to defend against drone warfare plays out in Ukraine. Proliferation of small, cheap drones is changing the way militaries think about air defense', *Wall Street Journal*, 15 December. Available at: www.wsj.com/world/the-race-to-defend-against-drone-warfare-plays-out-in-ukraine-96335409 (Accessed 2 June 2024).

Malakhov, V. (2018) 'Izobretenie traditsii kak mem i kak cherta sotsiokul'turnoi real'nosti: Erik Hobsbaum i ego soavtory v postsovetskom kontekste' ('The invention of tradition as a meme and as a feature of sociocultural reality: Eric Hobsbawm and his co-authors in the post-Soviet context'). *Neprikosnovenyyi, Zapas* 3 (119).

Mansour, R. (2023) 'Tackling Iraq's unaccountable state', Chatham House. Available at: www.chathamhouse.org/2023/12/tackling-iraqs-unaccountable-state?utm_source=linkedin.com&utm_medium=organic-social&utm_campaign=iraq&utm_content=democracy

Mate, A. (2003) 'Pillage is forbidden', *The Guardian*, 7 November. Available at: www.theguardian.com/world/2003/nov/07/iraq.comment

May, Robert E. (2002) *Manifest destiny's underworld*. Chapel Hill: University of North Carolina Press.

McCarthy, Rory (2001) *The Guardian*. Available at: www.theguardian.com/world/2001/dec/01/afghanistan.rorymccarthy1

McCartney, James (2015) *America's war machine: Vested interests, endless conflicts*. New York: Thomas Dunne Books.

McCoy, Alfred W. (1972) *The politics of heroin in Southeast Asia*. New York: Harper & Row.

McDonough, Siobhan (2005) 'Third columnist was paid by Bush agency', published by *Associated Press*, 28 January.

McGlynn, J. (2023) *Memory makers: The politics of the past in Putin's Russia*. London: Bloomsbury.

McTauge, T. (2022) 'What America's great unwinding would mean for the world', *The Atlantic*, 8 August. Available at: www.theatlantic.com/international/archive/2022/08/eur ope-america-military-empire-decline/670960/

Mearsheimer, John (2022) 'John Mearsheimer on why the West is principally responsible for the Ukrainian crisis', *The Economist*, March 19 [online] Available at: www.econom ist.com/by-invitation/2022/03/11/john-mearsheimer-on-why-the-west-is-principally-resp onsible-for-the-ukrainian-crisis

Memorial Platform (2023) Those killed in Russia's war against Ukraine. Available at: www.victims.memorial/people/dmytro-tsys

Memorial platform tweet, 16 May (2023) [online] Available at: https://twitter.com/memoria lua/status/1658522120538791936

Memorial platform tweet, 26 May (2023) [online] Available at: https://twitter.com/memoria lua/status/1661975610078511104

Memorial platform tweet, 3 June (2023) [online] Available at: https://twitter.com/memoria lua/status/1664874554055880705

Memorial platform tweet, 14 June (2023) [online] Available at: https://twitter.com/memoria lua/status/1669065075250831368

Memorial Platform tweet, 29 July (2023). Available at: https://twitter.com/memorialua/sta tus/1685305949177724928/photo/1

Micklethwait, John and Adrian Woolridge (2004) *The right nation*. New York: The Penguin Press.

Middle East Monitor (2024). 'Indonesia files lawsuit against Israel at IJC', 20 January. Available at: www.middleeastmonitor.com/20240120-indonesia-files-lawsuit-against-isr ael-at-icj/

Millan, Eva (2023) 'Historian Richard J. Evans: "Putin's goals are very ambitious; Hitler's had no limits"', *El Pais*, 18 August. Available at: https://english.elpais.com/culture/2023-08-18/historian-richard-j-evans-putins-goals-are-very-ambitious-hitlers-had-no-lim its.html

Ministry of Defence 'British fatalities. Operations in Iraq'. Available at: www.gov.uk/gov ernment/fields-of-operation/iraq

Misra, A (2004) *Afghanistan: The labyrinth of violence*. Cambridge: Polity.

Misra, A. (2002) "The Taliban, Radical Islam and Afghanistan," *Third World Quarterly*, 23(3). Pp. 577–589. Available at: https://www.tandfonline.com/doi/pdf/10.1080/014365 90220138349

Moller-Nielsen, Thomas (2022) 'Why did Russia launch this catastrophic war?', *Current Affairs*, 27 October. Available at: www.currentaffairs.org/2022/10/why-did-russia-lau nch-this-catastrophic-war

Monbiot, George (2001) 'America's pipe dream', *The Guardian*, 23 October. Available at: www.theguardian.com/world/2001/oct/23/afghanistan.terrorism11

Morgenthau, H.J. (1973) *Politics among nations: The struggle for power and peace*, 5th edition. New York: Alfred A Knopf.

Motil, Alexander J.(2022), 'Putin's Russia rose like Hitler's Germany – and could end the same', *The Hill*, 5 March. Available at: https://thehill.com/opinion/national-security/3470 515-putins-russia-rose-like-hitlers-germany-and-could-end-the-same/

Myre, Greg (2022) *NPR*, 12 March. Available at: www.npr.org/2022/03/12/1085861999/russias-wars-in-chechnya-offer-a-grim-warning-of-what-could-be-in-ukraine

Narea, N. (2023) 'How the US became Israel's closest ally', *Vox*, 13 October. Available at: www.vox.com/world-politics/23916266/us-israel-support-ally-gaza-war-aid

NATO (2023 i) 'NATO 2022 strategic Concept', *NATO*, 3 March. Available at: www.nato.int/cps/en/natohq/topics_210907.htm

NATO (2023) 'NATO Secretary General Jens Stoltenberg in a conversation on "The Road to Vilnius" at the Brussels Forum', *NATO*, 24 May. Available at: www.nato.int/cps/en/nat ohq/opinions_214799.htm

New York Times (2023) 'Troop deaths and injuries in Ukraine war near 500,000, U.S. officials say', 18 August. Available at: www.nytimes.com/2023/08/18/us/politics/ukraine-russia-war-casualties.html

New York Times (1998) 'On piping out Caspian oil, U.S. insists the cheaper, shorter way isn't better', 8 November. Available at: https://www.nytimes.com/1998/11/08/world/on-piping-out-caspian-oil-us-insists-the-cheaper-shorter-way-isn-t-better.html

noi (1998) 'PNAC letter to president', 26 January. Available at: https://noi.org/wp-content/uploads/2016/01/iraqclintonletter1998-01-26-Copy.pdf (Accessed 2 June 2024).

Notte, H. (2016) 'Russia in Chechnya and Syria: Pursuit of strategic goals', *Middle East Policy*, 23(1): 59–74.

Oakford, S. (2017) 'The United States used depleted uranium in Syria', *Foreign Policy*. Available at: https://foreignpolicy.com/2017/02/14/the-united-states-used-depleted-uranium-in-syria/

OHCHR (2022) 'UN Human Rights in Ukraine'. Available at: www.ohchr.org/en/countries/ukraine/our-presence

Open Society (2004) 'Disorder, negligence and mismanagement: How the CPA handles Iraq reconstruction funds', September. Available at: www.opensocietyfoundations.org/uplo ads/8749ed7c-83dc-4bf1-bb7c-e96bedf9c1ea/irwreport_20041001.pdf

Orend, Brian (2019) *War and political theory*. Cambridge: Polity.

Orozco, E.C. (1980) *Republican protestanism in Aztlán: The encounter between Mexicanism and Anglo-Saxon secular humanism in the United States Southwest*. Glendale: Peterins Press.

Pacheco, Nicholas P. (2022) 'How doctrine and delineation can help defeat drones'. War on the Rocks. *Texas National Security Review* [online] Available at: https://warontherocks.com/2022/12/how-doctrine-and-delineation-can-help-defeat-drones/

Pearce, Kimber Charles and Charles Pearce (2001) *Rostow, Kennedy, and the rhetoric of foreign aid*. East Lansing: Michigan State University Press.

Pierini, M. (2015) 'Assadland, a Russian protectorate', *Carnegie Endowment for International Peace*, 22 September. Available at: https://carnegieeurope.eu/2015/09/22/assadland-russian-protectorate-pub-61358 (Accessed 2 June 2024).

Piper, I. and Dyke, J. (2021) 'Tens of thousands of civilians likely killed by US in "Forever Wars"', *Airwars*. Available at: https://airwars.org/investigations/tens-of-thousands-of-civilians-likely-killed-by-us-in-forever-wars/

Policy Paper (2021) 'Global Britain in a competitive age: The integrated review of security, defence, development and foreign policy'. Available at: www.gov.uk/government/publi cations/global-britain-in-a-competitive-age-the-integrated-review-of-security-defence-development-and-foreign-policy/global-britain-in-a-competitive-age-the-integrated-rev iew-of-security-defence-development-and-foreign-policy

Politico (2022) 'What does Putin really want?', 25 February. Available at: www.politico.com/news/magazine/2022/02/25/putin-russia-ukraine-invasion-endgame-experts-00011652

Polsby, Nelson W. and Aaron Wildavsky (1991) *Presidential elections: Contemporary strategies of American Electoral Politics*, 8th edition. New York: The Free Press.

Puri, Samir (2022) *Russia's road to war with Ukraine*. Hull, UK: Biteback Publishing.

Putin, V. (2015) 'Speech at the 70th session of the UN Assembly'. Available at: www.europarl.europa.eu/meetdocs/2014_2019/documents/d-ru/dv/dru_20151015_06/dru_20151015_06en.pdf

Quinn, Ben (2014) 'Prince Charles reported to have likened Vladimir Putin to Hitler', *The Guardian*, 21 May. Available at: www.theguardian.com/uk-news/2014/may/21/prince-charles-reported-to-have-likened-putin-to-hitler

Rasheed, Zaheena (2023) 'How China became the world's leading exporter of combat drones', *Al Jazeera*, 24 January [online] Available at: www.aljazeera.com/news/2023/1/24/how-china-became-the-worlds-leading-exporter-of-combat-drones#:~:text=Data%20from%20the%20Stockholm%20International,exporter%20of%20the%20weaponised%20aircraft

Register of Russian War Criminals website. Available at: https://rwc.shtab.net/en/troopers

Reuters (2022) 'What are the Minsk agreements on the Ukraine conflict?', 21 February. Available at: www.reuters.com/world/europe/what-are-minsk-agreements-ukraine-conflict-2022-02-21/ (Accessed 2 June 2024).

Reuters (2024) 'At ICJ, South Africa accuses Israel of genocide in Gaza', 11 11 January. Available at: www.reuters.com/world/middle-east/israel-safrica-face-off-un-top-court-gaza-genocide-case-2024-01-11/ (Accessed 2 June 2024).

Reynolds, David (2016) *Afghanistan: Britain's War in Helmand*. Plymouth: DRA Publishing.

Rogers, Paul (2021) *Losing control: Global security in the twenty-first century*, 4th edition. London: Pluto Press.

Rogers, Paul (2022) 'Ukraine and global human security', *Rethinking Security*, 31 May. Available at: https://rethinkingsecurity.org.uk/2022/05/31/ukraine-and-global-human-security/

Rogers, Paul (2023) 'Our global culture of war means guaranteed profits for the arms industry', *openDemocracy*, 23 June. Available at: www.opendemocracy.net/en/arms-industry-shareholder-capitalism-perfect-war-syria-iraq-ukraine/

Rohde, D. (2012) 'The Drone Wars', *Reuters*. Available at: www.reuters.com/article/us-david-rohde-drone-wars-idUSTRE80P11I20120126 (Accessed 2 June 2024).

Rundle, Guy (2001) 'The New World Order under siege', *Arena Magazine*, October–November. P. 2

Russell, R. (2022) 'Ukraine war: Russian soldier jailed for life after murdering unarmed Ukrainian civilian in first war crimes trial', *Sky News*, 23 May. Available at: https://news.sky.com/story/ukraine-war-russian-soldier-21-jailed-for-life-after-murdering-unarmed-ukrainian-civilian-in-first-war-crimes-trial-12619520

Saied, A.A., Ahmed, S.K., Metwally, A.A. and Aiash, H. (2023) 'Iraq's mental health crisis: A way forward?', *The Lancet*, 402(10409): 1235–1236. Available at: www.thelancet.com/journals/lancet/article/PIIS0140-6736(23)01283-7/fulltext

Samuels, B. (2023) 'ICC prosecutor visits Rafah crossing on Egypt-Gaza border amid Israeli war crimes allegations', *Haaretz*, 29 October. Available at: www.haaretz.com/israel-news/2023-10-29/ty-article/.premium/icc-prosecutor-visits-egypt-gaza-border-amid-israeli-war-crimes-allegations/0000018b-7c48-d51e-a3cb-7d6c18860000

Sandole, Dennis J.D.(2005) 'The Western-Islamic "clash of civilisations": The inadvertent contribution of the Bush presidency', *Peace and Conflict Studies Journal*, 2(2). Available at: https://nsuworks.nova.edu/pcs/vol12/iss2/2/

Saul, J. (2023) 'The hidden trauma of moral injury', *Psychotherapy Networker*. Available at: www.psychotherapynetworker.org/article/hidden-trauma-moral-injury

Savage, Charlie, Schmitt, Eric, Khan, Azmat, Hill, Evan and Koettl, Christoph (2022), *New York Times*. Available at: www.nytimes.com/2022/01/19/us/politics/afghanistan-drone-strike-video.html (Accessed 2 June 2024).

Savell, Stephanie (2023). 'How death outlives war: The reverberating impact of the post-9/11 wars on human health', *Watson Institute-Brown University*, 15 May. Available at: https://watson.brown.edu/costsofwar/files/cow/imce/papers/2023/Indirect%20Deaths.pdf

Schecter, A. and Simmons, K. (2023) 'Israel's secret air war in Gaza and the West Bank', *NBC News*, 21 November. Available at: www.nbcnews.com/investigations/israels-secret-air-war-gaza-west-bank-rcna126096

Schlesinger, Arthur M. (1967) *The bitter heritage: Vietnam and American democracy 1941–1966.* New York: Houghton Mifflin Co.

SCM (2023) 'French magistrates issue arrest warrants for the Syrian president and three associates for chemical weapons attacks'. Available at: https://scm.bz/en/french-magistrates-issue-arrest-warrants-for-the-syrian-president-and-three-associates-for-chemical-weapons-attacks/

Searcey, Dionne (2021) 'Boko Haram is back. With better drones'. *New York Times*, September 23 [online] Available at: www.nytimes.com/2019/09/13/world/africa/nigeria-boko-haram.html (Accessed 2 June 2024).

Shlapentokh, Dmitry (2009) 'Review of Marlene Laruelee's Russian Eurasianism: An ideology of empire', *Revue D'etudes Comparatives Est-Ouest*, 40: 197–198. Available at: www.cairn.info/revue-d-etudes-comparatives-est-ouest1-2009-2-page-197.htm (Accessed 2 June 2024).

SIGAR (Special Inspector General for Afghanistan Reconstruction) (2021) What we need to learn: Lessons from twenty years of Afghanistan reconstruction, *SIGAR*, August. Available at: www.sigar.mil/pdf/lessonslearned/SIGAR-21-46-LL.pdf

SIPRI (Stockholm International Peace Research Institute) (2022) The top 15 military spenders . Available at: www.sipri.org/visualizations/2023/top-15-military-spenders-2022

Skocpol, Theda and Morris P. Fiorina. (1999) 'Making sense of the civil engagement debate', in Theda Skocpol and Morris P. Fiorina, eds. *Civic engagement in American democracy*. Washington: Brookings Institution Press: 1–23.

Sly, L. (2023) '66,000 war crimes have been reported in Ukraine. It vows to prosecute them all', *The Washington Post*, 6 February. Available at: www.washingtonpost.com/world/2023/01/29/war-crimes-ukraine-prosecution/

Smith, David (2023) *Endless holocausts.* New York: Monthly Review Press.

Smith, Erin (2021) https://medium.com/@erin.smith_2213/the-9-11-firefighter-whose-name-you-should-know-c33d49771074

Smith, N. and Overton, I. (2022) 'More serving British service personnel have killed themselves since 1984 than have died in combat, AOAV research finds', Action on Armed Violence. Available at: https://aoav.org.uk/2022/more-serving-british-service-personnel-have-killed-themselves-since-1984-than-have-died-in-combat-aoav-research-finds/

Smith, Patrick (2022) 'Putin compares himself to Peter the Great in fight to expand Russia', *NBC News*, 10 June. Available at: www.nbcnews.com/news/world/putin-ukraine-russia-tsar-peter-great-imperialism-rcna32909

Soteriou, E. (2022) 'Russia poses "most immediate threat" to UK security, says ex MI6 chief, *LBC*, 13 September,. Available at: www.lbc.co.uk/news/russia-immediate-thr eat-uk-security-ex-mi6-chief/#:~:text=%22The%20most%20immediate%20threat%20 to,a%20conflict%20between%20the%20two

Special Inspector General for Iraq Reconstruction (2005) 'Report to Congress', 30 January. Available at: https://apps.dtic.mil/sti/pdfs/ADA489361.pdf

Stanford (2007) 'Courting disaster: The fight for oil, water and a healthy planet', 13 October. Available at: https://fsi.stanford.edu/events/courting_disaster_the_fight_for_oil_water_a nd_a_healthy_planet

Stohl, Rachel and Grillot, Suzette (2009) *The international arms trade*. Cambridge: Polity.

Strang, A.B. and Quinn, N. (2019). 'Integration or isolation? Refugees' social connections and wellbeing', *Journal of Refugee Studies*, 34(1): 328–353.

Strawser, B.J. (2012) 'The morality of drone warfare revisited', *The Guardian*, August 6. Available at: www.theguardian.com/commentisfree/2012/aug/06/morality-drone-warf are-revisited#:~:text=The%20Guardian%20has%20graciously%20offered,their%20op erators%20to%20greater%20risk.

Suganami, Hidemi (1996) *On the causes of war*. Oxford: Clarendon Press.

Suitt, T.H. (2021) 'High suicide rates among United States service members and veterans of the post9/11 wars', Watson Institute. Available at: https://watson.brown.edu/costsof war/files/cow/imce/papers/2021/Suitt_Suicides_Costs%20of%20War_June%2021%202 021.pdf

Swain, Judith L. (2008) 'Our United States legacy: Lessons learned from the British Empire', *Journal of Clinical Investigation*, 118(11): 3802–3804. Available at: www.ncbi.nlm.nih. gov/pmc/articles/PMC2575706/

Tawfeeq, Mohammed (2021) 'Iraq estimates that $150 billion of its oil money has been stolen from the country since the US-led invasion of 2003', *CNN*, 23 May. Available at: https://edition.cnn.com/2021/05/23/middleeast/iraq-oil-money-us-invastion-intl/ index.html

The Guardian (2022) 'Scores of Iraqis injured in anti-government protests in Baghdad', 1 October. Available at: www.theguardian.com/world/2022/oct/01/scores-of-iraqis-injured- in-anti-government-protests-in-baghdad

The Guardian (2024) 'Russia-Ukraine war', 6 January. Available at: www.theguardian.com/ world/live/2024/jan/06/russia-ukraine-war-aid-white-house-putin-zelenskiy

The Nation The deaths of Afghans database (n.d.) Available at: www.thenation.com/afgh anistan-database/

Thorn, T. (2013) 'Revisiting the Iraq war, international policy digest'. Available at: https:// intpolicydigest.org/revisiting-the-iraq-war/

Todd, Emmanuel (2003) 'Interview with Emmanuel Todd', *Prospect Magazine*, 19 June. Available at: www.prospectmagazine.co.uk/essays/59335/emmanuel-todd

Torelli, C. (2023) 'Suffer the children: The child casualties of Operation Swords of Iron', Action on Armed Violence. Available at: https://aoav.org.uk/2023/suffer-the-children-the- child-casualties-of-operation-swords-of-iron/

Treisman, D. (2018) *The new autocracy: Information, politics, and policy in Putin's Russia*. Washington: Brookings Institution.

Turner, Frederick Jackson (1996) *The frontier in American history*. Mineola: Dover Publications.

UNAMA (2019) 'Afghanistan protection of civilians in armed conflict annual report: 2018'. Available at: https://reliefweb.int/report/afghanistan/afghanistan-protection-civilians- armed-conflict-annual-report-2018-endarips

UNAMA reports (n.d.) https://unama.unmissions.org/protection-of-civilians-reports.

UNAMI/OHCHR (2020) 'Human rights violations and abuses in the context of demonstrations in Iraq, October 2019 to April 2020'. Available at: www.ohchr.org/sites/default/files/Documents/Countries/IQ/Demonstrations-Iraq-UNAMI-OHCHR-report.pdf

UNHCR (2023) 'Ukraine emergency'. Available at: www.unrefugees.org/emergencies/ukraine/#:~:text=More%20than%206.2%20million%20refugees,(as%20of%20July%202023).&text=Approximately%2017.6%20million%20people%20are%20in%20need%20of%20humanitarian%20assistance%20in%202023

UNHR. (2022). *Ukraine: Behind the numbers.* Kyiv: United Nations.

UNHR. (2023). Ukraine: civilian casualty update. Kyiv: United Nations Human Rights, Office of The High Commissioner.

United Nations (UN) (1948). 'Convention on the prevention and punishment of the crime of genocide', 9 December. Available at: www.un.org/en/genocideprevention/documents/atrocity-crimes/Doc.1_Convention%20on%20the%20Prevention%20and%20Punishment%20of%20the%20Crime%20of%20Genocide.pdf

United Nations (UN) (2023). 'Press release', 24 October. Available at: https://press.un.org/en/2023/sgsm22003.doc.htm

United Nations (n.d.) 'War crimes', Office on Genocide Prevention and the Responsibility to Protect. Available at: www.un.org/en/genocideprevention/war-crimes.shtml

United Nations Human Rights Council (2023) 'Impact of casualty recording on the promotion and protection of human rights', Report of the United Nations High Commissioner for Human Rights. Available at: https://reliefweb.int/report/world/impact-casualty-recording-promotion-and-protection-human-rights-report-united-nations-high-commissioner-human-rights-ahrc5348-enarruzh

United Nations International Criminal Tribunal for the Former Yugoslavia (n.d.), Slobodan Milošević Trial. Available at: www.icty.org/en/content/slobodan-milo%C5%A1evi%C4%87-trial-prosecutions-case

United Nations Security Council (2001) 'Resolution 1368: Adopted by the Security Council at its 4370th meeting'. Available at: https://ofac.treasury.gov/media/5651/download?inline

United States Department of State (2021) 'Iraq 2021 human rights report'. Available at: www.state.gov/wp-content/uploads/2022/03/313615_IRAQ-2021-HUMAN-RIGHTS-REPORT.pdf (Accessed 2 June 2024).

UN Office of the High Commissioner for Human rights (2023) 'Ukraine: Civilian casualties as of 8 October 2023'. Available at: https://reliefweb.int/report/ukraine/ukraine-civilian-casualties-8-october-2023-enruuk#:~:text=Civilian%20casualties%20from%201%20to,sex%20is%20not%20yet%20known

Uppsala Conflict Data Program (n.d.) https://ucdp.uu.se/

Uppsala Conflict Data Program (2023) 'UCDP methodology'. Available at: https://www.uu.se/en/department/peace-and-conflict-research/research/ucdp/ucdp-methodology

Varma, T. and Huggard, K. (2023) 'France responds to the Israel-Gaza crisis', Brookings, 15 December. Available at: www.brookings.edu/articles/france-responds-to-the-israel-gaza-crisis/

Vincent, Faustine (2023) 'The secret carnage of military losses in Ukraine', *Le Monde*, 24 August. Available at: www.lemonde.fr/en/international/article/2023/08/24/behind-the-secrecy-of-military-losses-in-ukraine-carnage-on-a-massive-scale_6107230_4.html

Ward, E. (2023) 'Putting names to numbers. The creation of a systematic casualty recording database for the ongoing Russian aggression within Ukraine', *Journal of Global*

Faultlines, 10(2): 238–251. Available at: https://www.scienceopen.com/hosted-docum ent?doi=10.13169/jglobfaul.10.2.0238

Warsi, F. (2023) 'How the US and UK tried to justify the invasion of Iraq', *Al Jazeera*, 19 March. Available at: www.aljazeera.com/news/2023/3/19/examining-justifications-us-invasion-iraq

Washington Post (2023) 'Putin plans for a long struggle in Ukraine. The U.S. needs to do the same', *Washington Post Opinion*, 23 August. Available at: www.washingtonpost.com/opinions/2023/08/23/ukraine-18-months-aid-war/

Watson institute (2021) 'Millions displaced by U.S. post-9/11 wars'. Available at: https://wat son.brown.edu/costsofwar/files/cow/imce/papers/2021/Costs%20of%20War_Vine%20 et%20al_Displacement%20Update%20August%202021.pdf

Watson Institute (2021) 'U.S. and allied killed'. Available at: https://watson.brown.edu/cos tsofwar/costs/human/military/killed

Watson Institute (2023) 'Figures'. Available at: https://watson.brown.edu/costsofwar/figu res/2021/WarDeathToll)

Watson Institute (2023) 'Human costs'. Available at: https://watson.brown.edu/costsofwar/ costs/human

Weber, M. (1994) 'The profession and vocation of politics', in Lassman, P. and Speirs, R., eds. *Weber – political writings.* Cambridge: Cambridge University Press: 309–369.

White House (2021a) 'Remarks by President Biden on the way forward in Afghanistan', 14 April. Available at: www.whitehouse.gov/briefing-room/speeches-remarks/2021/04/14/ remarks-by-president-biden-on-the-way-forward-in-afghanistan/#:~:text=We'll%20conti nue%20to%20support,over%20the%20past%20two%20decades.

White House (2021b) 'Remarks by President Biden on the Drawdown of U.S. Forces in Afghanistan', 8 July. Available at: https://www.whitehouse.gov/briefing-room/speec hes-remarks/2021/07/08/remarks-by-president-biden-on-the-drawdown-of-u-s-forces-in- afghanistan/

Whyte, Dave (2007) 'The crimes of neoliberal rule in occupied Iraq', *The British Journal of Criminology*, 47(2). Available at: https://academic.oup.com/bjc/article/47/2/177/519 163?login=true

Williams, H. (2012) *Kant and the end of war.* Palgrave Macmillan.

Wolf, Martin (2017) 'The long and painful journey to world disorder', *Financial Times*, 5 January. Available at: www.ft.com/content/ef13e61a-ccec-11e6-b8ce-b9c03770f8b1

Wolf, Martin (2018). 'Davos 2018: The liberal international order is sick', *Financial Times*, 23 January. Available at: www.ft.com/content/c45acec8-fd35-11e7-9b32-d7d59aace167

Woods, C (2016) 'Why White House civilian casualty figures on civilian are a wild under-estimate'. *The Bureau of Investigative Journalism*. Available at: www.thebureauinvestiga tes.com/opinion/2016-07-01/comment-official-estimates-show-civilians-more-likely-to-be-killed-by-cia-drones-than-by-us-air-force-actions-the-reality-is-likely-far-worse/ #:~:text=The%20US%20claims%20that%20between,than%20in%20conventional%20 US%20airstrikes.

Woodward, B. (2023) 'The UK supports Israel's right to defend itself against Hamas but Israel must be targeted in achieving that goal: UK statement at the UN Security Council', GOV.UK. Available at: www.gov.uk/government/speeches/the-uk-supports-israels-right-to-defend-itself-against-hamas-but-israel-must-be-targeted-in-achieving-that-goal-uk-statement-at-the-un-security-cou#:~:text=The%20UK%20continues%20to%20supp ort,precise%20in%20achieving%20that%20goal

Zakharchenko, K. (2023) 'Ukraine has identified over 180,000 war criminals and created a registry', *Kyiv Post*, 15 November. Available at: www.kyivpost.com/post/24162

Zelensky, V. (2022) *A message from Ukraine*. New York: Hutchinson Heinemann, Penguin Random House.

Zerubavel, Y. (2005) *Recovered roots*. Chicago: University of Chicago Press.

Zygar, Mikhail (2023) 'The man behind Putin's warped view of history', *The New York Times*, 19 September. Available at: www.nytimes.com/2023/09/19/opinion/putin-russia-medinsky.html

Index

Note: Page numbers in **bold** refers to Tables. Endnotes are indicated by the page number followed by "n" and the note number e.g., 87n2 refers to note 2 on page 87.

Action on Armed Violence (AOAV) 1–3, 92–94
Afghanistan: bombing on 3; civilians killing 7; coalition killings of civilians **37**, 38; deliberate detention operations (DDO) in 58; Islamic law in 16; opium production 19; Taliban in 16; US committed to withdrawing troops 19; US invasion of 11, 14–19, 28, 34–39
air strikes 11; invasion of Iraq 32; against ISIS 32–33; War on Terror 32–33
Airwars 9, 32–33, 48, 57, 113
Al-Assad, B. 79, 111
Al Mahdi, A. A. F. 107–108
al-Qaeda 34, 36, 50, 105
Anti-Ballistic Missile Treaty 98
armed drones, victims of 63; *see also* drone strikes
Army of Drones 6
artificial intelligence 101, 106
artificial intelligence (AI) arms race 5
Ashura Brigades 5
AUMF *see* Authorization for Use of Military Force (AUMF)
Authorization for Use of Military Force (AUMF) 11, 36
autonomous weapons systems (AWS) 5

Ba'athists 50
Baghdad: bodies found in 2006 **46**; bombing of 20
bin Laden, O. 35
biological weapons 14
Blair, T. 25
blast at Kabul airport 39

BRICS+ bloc 72
Brown University Cost of War programme 33
Brzezinski, Z. 14–16, 98
Bush, G. W. 17, 20, 23, 82

chemical, biological, radiological and nuclear agents (CBRN) 49
chemical warfare 14
Children of War 92, 94–95
China, mass detention of Uyghur people in Xinjiang province 115
civilian fatalities 1–3, 8–9; 2003–2013 **42**; child casualties 3–4; in Iraq 45; ISIS years (2014–2017) **43**
clash of civilisations 14
Clausewitz, C. v. 67
Claw-Lock 113
Coalition Provisional Authority (CPA) 20–21
Cold War 70–71; aggression 79
collateral damage 4, 32, 34
Commission on Human Security (CHS) 8
Commonwealth of Independent States (CIS) 13
community insecurity 68
community security 33, 49
Correlates of War project 33
Crimea 91
crimes: of aggression 107; committed by British 60; against humanity 107
cyber-terrorism 14

deliberate detention operations (DDO) in Afghanistan 58
depleted uranium (DU) 49

Development Fund for Iraq (DFI) 20
Devil's Decade 69
Donbas 91
drone strikes 2, 11, 33, 38, 64, 103,
    105–106; against Houthi forces 2;
    in Pakistan 33; in Somalia 33; by
    state actors 2–3; targeting militant
    organisations 33; in Yemen 33

East–West Cold War 115
economic insecurity 68
economic security 33, 49
ethics of self-defence 97
ethnic cleansing 13
EU: economic strength measured against 28;
    joint declaration of support for Ukraine 5;
    policies 76; Serbia peace plan with 117
Eurasianism 83, 87n2
European Asylum Support Office 60
European Centre for Constitutional and
    Human Rights (ECCHR) 59

financial meltdown of 2008 1
food security 49
Forbes, S. 23
Foreign Islamist volunteers 50
Fukuyama, F. 21–23

Geneva Conventions 57–58
genocide 107
Genocide Convention 126
global hegemony 29–30
Global Terrorism Database (GTD) 33, 48
Gramsci, A. 27, 119–120

health security 49
high- value target (HVT) cell 41
'hub-and-spokes' arrangement 28
human casualties of War on Terror 9;
    casualty recording 57–61; contextualising
    death and human suffering 61–66; human
    rights 57–61; human security 48–54,
    68; international law 57–61; invasion of
    Afghanistan 34–39; invasion of Iraq
    39–48; mainstream/ traditional perspective
    66–68; post-traumatic stress disorder
    54–57
Human Rights Monitoring Mission in
    Ukraine (HRMMU) 92–93
Human Rights Watch 114
human security 48–54; categories of 8;
    crises 1–2, 8–10
human (in)security 1–7, 68

International Centre for the Prosecution of
    the Crime of Aggression against Ukraine
    (ICPA) 109
International Court of Justice (ICJ) 124,
    126–127
International Criminal Court (ICC) 59, 88,
    125
International Criminal Tribunal for the
    Former Yugoslavia 109
international humanitarian law (IHL)
    principles 5, 92
International Stabilisation Assistance Force
    (ISAF) 34
Iraq: air strikes in 32; bankrupt economy
    21; neoliberal utopia 20; oil income 20;
    US-UK coalition invasion of 19–22, 32,
    40–41; weapons of mass destructions
    (WMD) 19
Iraq Body Count (IBC) 9, 20, 33–34, 40,
    40, 41, 61, 68, 110; incident a5497 **64**;
    incident a5885 **44**; incident a6079 **65**;
    incident a6250 **44**; incident k001 **64**;
    incident k1019 **65**; incident k4435 **45**
ISIS 9, 38; air strikes against 9, 32–33;
    civilian deaths during war against 33;
    victim 44
Israel-Hamas war 68, 119
Israeli air strike in Gaza 3

Joint Multinational Training Group-Ukraine
    99
just war theory 35, 96; principles of 97

'kamikaze' quadcopter drones 106
Kharkiv 103
KLA in Kosovo 14
Kremlin 78

laws of armed conflict 94–95

mass killing of Iraqis 39–40
Memorial Platform 92–94
mental health crisis 54
mental illnesses 56
militarism 117
military-industrial complex 102,
    121–122
Minsk Agreements 75; Minsk-1 deal 75;
    Minsk-2 deal 75–76
moral injury 55–56

NATO 10, 71, 73, 99, 104–105, 113
Netanyahu, B. 111, 125

New Cold War 119–127; multi-level conflicts 120; *see also* Russian invasion of Ukraine

Obama, B. 21, 50, 57
oil and gas transportation routes 14
Operation Enduring Freedom 35
Operation Freedom's Sentinel 35
Operation Iraqi Freedom 11, 27, 39–41, 52
Operation Pillar of Defense 3
Operation Protective Edge 3
Operation Swords of Iron 2–3
Operation Wall Guardian 3

Palestinian civilians with the surname Husuna **66**
Palestinian Islamic Jihad 105
Palmer, Orio 33–34
Pax Americana 22, 30
PKK (Kurdish terrorist group) 105
policy of Courageous Restraint 38
political security 33, 49
post-Cold War US foreign policy 13
post-traumatic stress disorder 33, 54–57
precision strikes 44
principle of unnecessary risk 64
Project for the New American Century (PNAC) 22–23, 35, 83
Putin, V. 10, 72, 83, 125; ICC issued warrants of arrest 25, 125; *see also* New Cold War; Russian invasion of Ukraine

Reagan Doctrine, 1985 15
Red Army soldiers 78
Red Cross 49, 57
Register of Russian War Criminals 114
remote warfare 10, 88
rules-based international order 119
Russian annexation of Crimea 91
Russian invasion of Ukraine 1, 10, 69–77, 101, 119–123; Izborsky Club *(Izborskii klub)* 83–84; Minsk (Minsk- 1) negotiations 75; politics of the past 77–80; reason for 80–82; war damage and casualties 84–86
Russophobia 79, 84

Salafi/Wahhabi 'jihadists' 50
Saudi-led coalition air strikes 2
self-defensive killing 97–98
September 11 (9/11) attacks 1, 11–14
Shishimarin, V. 10, 88–89

SIGACTS (Significant Activities) reports 45
Soviet power, collapse of 12–13
Special Inspector General for Afghanistan Reconstruction (SIGAR) 18
Special Inspector General for Iraq Reconstruction (SIGIR) 20
Stockholm International Peace Research Institute 102
Syrian Center for Media and Freedom of Expression (SCM) 112
Syrian National Army (SNA) 113

Taliban 15, 36
Turkish Military Operations in Northern Iraq 113

Ukraine: refugees from 5; war in 5–7; *see also* Russian invasion of Ukraine
UNAMA 33, 36
United Nations Security Council 35–36
Universal Declaration of Human Rights 68
unmanned aerial systems (UAS) 99, 105
UN's Genocide Convention 124
Uppsala Conflict Data Program (UCDP) 33, 47
US: arms exporter 102; foreign policy 13; invasion of Afghanistan 11, 14–19; troops killed at Kabul airport **38**; *see also* War on Terror
US/NATO remote war 101, 104
US–UK coalition invasion of Iraq 19–22, 26
US War on Terror *see* War on Terror

violations of international humanitarian laws 58
violations of the Genocide Convention 124

war crimes 10, 58, 88–92, 107; casualties, recording 92–95; and ethics 95–101; global security dynamics 114–118; international justice 107–114; remote warfare in Ukraine 101–107; Ukraine's just war 95–101
war damage and casualties 84–86
War on Terror 11, 17, 27, 119; Afghanistan 14–19; air strikes 32–33; conflict 57; Invasion of Iraq, March 2003 19–22, 40–41; mass bombing campaigns 32; New American Century 22–25; oil and

25–27; September 11 1, 11–14; US world hegemony 27–31
weapons of mass destruction (WMD) 19, 22
WikiLeaks 33
Wing Loong II drones 105
World Trade Centre 17

Yazidis 63
Yekatom, A. 108
Yeltsin, B. 72, 82
Yugoslavia, war in 71–72

Zelensky, V. 89, 91, 100, 103

For Product Safety Concerns and Information please contact our EU
representative GPSR@taylorandfrancis.com
Taylor & Francis Verlag GmbH, Kaufingerstraße 24, 80331 München, Germany

www.ingramcontent.com/pod-product-compliance
Lightning Source LLC
Chambersburg PA
CBHW060314220326
41598CB00027B/4325

*9 7 8 1 0 3 2 5 4 0 3 6 8 *